ROUTLEDGE LIBRARY EDITIONS:
INTERNATIONAL SECURITY STUDIES

Volume 13

MILITARY APPLICATIONS OF MODELING

MILITARY APPLICATIONS OF MODELING

Selected Case Studies

FRANCIS P. HOEBER

LONDON AND NEW YORK

First published in 1981 by Gordon and Breach Science Publishers

This edition first published in 2021
by Routledge
2 Park Square, Milton Park, Abingdon, Oxon OX14 4RN

and by Routledge
52 Vanderbilt Avenue, New York, NY 10017

Routledge is an imprint of the Taylor & Francis Group, an informa business

British Library Cataloguing in Publication Data
A catalogue record for this book is available from the British Library

ISBN: 978-0-367-68499-0 (Set)
ISBN: 978-1-00-316169-1 (Set) (ebk)
ISBN: 978-0-367-71242-6 (Volume 13) (hbk)
ISBN: 978-0-367-71246-4 (Volume 13) (pbk)
ISBN: 978-1-00-314998-9 (Volume 13) (ebk)

Publisher's Note
The publisher has gone to great lengths to ensure the quality of this reprint but points out that some imperfections in the original copies may be apparent.

Disclaimer
The publisher has made every effort to trace copyright holders and would welcome correspondence from those they have been unable to trace.

Military Applications of Modeling: Selected Case Studies

Francis P. Hoeber

Hoeber Corp.
Arlington, Virginia

GORDON AND BREACH SCIENCE PUBLISHERS
New York London Paris

Gordon and Breach, Science Publishers, Inc.
One Park Avenue
New York, NY 10016

Gordon and Breach Science Publishers Ltd.
42 William IV Street
London WC2N 4DE

Gordon & Breach
7-9 rue Emile Dubois
F-75014 Paris

Library of Congress Cataloging in Publication Data

Hoeber, Francis P.
Military applications of modeling.

(Military operations research, ISSN 0275-5823; v. 1)
Bibliography: p.
Includes index.
1. Military art and science—Mathematical models—Case studies. 2. Operations research—Case studies. 3. Mathematical models—Case studies. I. Title. II. Series.
U104.H56 355'.001'51 81-6446
ISBN 0-677-05840-3 AACR2

Library of Congress catalog card number 80-2744 ISBN 0 677 05840-3 ISSN: 0275-5823.

Editor's Introduction

General Sun Tzu, a Fifth-Century B.C. approximate contemporary of Confucius, is probably the first recorded practitioner of military operations research. In his epochal text, *The Art of War,* he observed:

> Know the enemy and know yourself; in a hundred battles you will never be in peril. When you are ignorant of the enemy but know yourself, your chances of winning or losing are equal. If ignorant both of your enemy and of yourself, you are certain in every battle to be in peril.

His advice has met the test of nearly 24 centuries. The complexity of the commander's problems has increased considerably since the days of the Chou Dynasty (in terms of battlefield area, perhaps 82 db), but remains qualitatively the same.

The essence of *knowing* is indeed analysis. No war is ever fought twice; the essence of preparation, of planning, is the analysis of data which are at best only partially relevant: historical data, exercise and test results, and intelligence. Such analysis and implicit extrapolation from the available quantitative information provide the basis for the planning of national defense: systems, strategies, and operations. This is military operations research.

The application of the disciplines now known collectively as military operations research is as old as mathematics itself. Archimedes was probably the first Government consultant on military operations, and DaVinci perhaps the first system engineer. The *sector,* the first analog calculator, reputedly fashioned by Galileo circa 1590, regularly included special scales for the design of fortifications. More recently, one of the first formal, testable bodies of theory of military engagements is that of F. W. Lanchester who in 1914 published a mathematical model of tactical warfare, embodying and quantifying the principle of concentration of forces, with emphasis on the then-new problem of the air battle.

During World War II the *corpus* of modern military operations research (and indeed the probable first use of the phrase "operations research") was seeded by the development of tactics utilizing the new radar machines, and by the successful optimization of antisubmarine warfare tactics developed by the U.S. Navy during World War II. This work was subsequently reported by Philip Morse and George Kimball, who published the first definitive text on the subject.

The art of military operations research has broadened and deepened since these beginnings, and its impact on systems planning, *inter alia,* has grown as well. As national defense grows more complex, so do the formal methods for its analysis.

Recognizing the criticality of analysis to the management of national defense and to the planning of warfare, and recognizing the need for a comprehensive body of professional literature in this discipline, the Military Operations Research Society is publishing this series. *This volume, like the others, represents the findings and opinions of its author, and does not represent any policy or position of the Editors, of MORS or of the United States Government.*

The present volume, by Francis P. Hoeber, offers a critical review of the techniques of mathematical modeling and their appropriate application, candidly presented by a master. The virtues of sophistication via simplicity, and the beauty of the artful finesse, emerge as the signature of successful modeling.

STEPHEN W. LEIBHOLZ

The Military Operations Research Society

This series of monographs represents one of the professional activities of the Military Operations Research Society, the professional society of the community of practitioners of operations research in the context of national security.

The purpose of the Military Operations Research Society is to enhance the quality and effectiveness of military operations research. To accomplish this purpose, the Society provides media for professional exchange and peer criticism among students, theoreticians, practitioners, and users of military operations research. These media consist primarily of the semiannual MORS symposia, their published proceedings, and special-purpose monographs. The forum provided by these media is directed to display the state of the art, to encourage consistent professional quality, to stimulate communication and interaction between practitioners and users, and to foster the interest and development of students of operations research. In performing its function, the Military Operations Research Society does not make or advocate official policy nor does it attempt to influence the formulation of policy. Matters discussed or statements made during the course of its symposia or printed in its publications represent the positions of the individual participants and authors, and not of the Society.

The Military Operations Research Society is operated by a Board of Directors consisting of 28 members, each of whom is elected by vote of the Board to serve a term of four years. Nominees for this election are individuals who have attained recognition and prominence in the field of military operations research and who have demonstrated an active interest in its programs and activities. Since a major portion of the Society's affairs is connected with classified services to military sponsors, the Society does not have a general membership in the sense that other professional societies have them. In place of this, the Society maintains a general distribution list of its clientele to whom announcements, newsletters, and information are routinely sent.

The Symposia and publications of the Society exhibit many different and personal viewpoints. Indeed, dissent is encouraged as a means of filtering, testing and refining the professional skills of its members and the soundness of their work. Consequently these products, of which the present volume is a part, represent the views of their authors, and not of the Society or the United States Government.

Preface

These case studies were designed for use in a graduate course on the Military Applications of Modeling in the Strategic and Tactical Sciences Program of the Air Force Institute of Technology (AFIT). They were planned to supplement, not replace, a textbook.

Five cases were prepared under contract number F33601-78-C-0075, dated 3 March 1978, between the Air Force Institute of Technology and the General Research Corporation, to which the author is a consultant. A draft version was used in the first session of the program, in the first quarter of 1979, and a final version was published by General Research Corporation in August 1979, revised with the benefit of the thoughtful criticisms and suggestions of the AFIT faculty and students who used the Casebook.

At AFIT's request, three additional cases were prepared by Hoeber Corp. under contract number F33600-79-C-0525, dated 11 September 1979. These cases (Chapter V, VII and VIII of the present volume) were again most helpfully critiqued by AFIT faculty and students after the first-quarter 1980 session.

The author wishes to thank a considerable number of people for their contributions, while leaving all save himself blameless for any errors:

At AFIT, Lt. Col. Saul Young, who taught the course both years. Saul Young's guidance was invaluable, and he was a pleasure to work with. The term papers of his students added many useful insights. Those of Lt. Col. Gregory A. Keethler and Major Thomas P. Flanagan on Chapter V should be especially noted.

At General Research Corporation, Phillip H. Lowry, for the first draft of Chapter VI, Jerome C. Ford, for assistance on Chapter II, Caroll J. Keyfauver and Howard S. Greer, for assistance on Chapter III, and James L.

Jones, Robert A. Gessert and Phillip H. Lowry, for many useful comments and suggestions throughout.

At Air Force Studies and Analysis, Lt. Col. Joseph D. Bester, for information for and comments on Chapter IV, Colonel John Friel, for his patient assistance with Chapter V, and Captain Gregory Tsoucallas, for extensive comments on Chapter VII.

At Hoeber Corporation, Robert M. Dannenberg, for many long hours spent in finding and checking sources for his patient efforts to keep the author honest in his quotations, logic and grammer.

Especially at MORS, Col. Charles F. ("Frank") Pilley, Jr., of Air Force Studies and Analyses, for his endless patience in heading the MORS review of this monograph and for reviewing every page himself, Judith Leibholz, for the tiresome task of making an index, MORS reviewers Lt. Col. R. S. "Hap" Miller, Carl Jones, Capt. Wayne Hughes, at the Naval Post Graduate School, Carlton Thorne at the Arms Control and Disarmament Agency, for patient reading and many helpful suggestions, and finally Stephen W. Leibholz, President of Analytics, for his work as editor and for the labor of supervising publication.

FRANCIS P. HOEBER

Contents

List of Figures

CHAPTER I

Introduction

THE NATURE OF MODELING

Military organizations have been the source of much of the development of modern, sophisticated modeling technique. One could take World War II as an arbitrary but not unreasonable historical starting point, and perhaps specifically the early British "operational research" on such problems as the operational use of radar and the design of aircraft search patterns for submarines in the North Atlantic.† But the concept of models and modeling is neither new nor specific to military applications.

The Greeks had highly abstract models of the nature of the universe, e.g.: the earth-fire-water-air and atomic models of the substance of things; the Euclidean geometry, the axioms of which were generally accepted as consistent with the real world; and the Ptolemaic geocentric model of the universe. Every artistic, scientific or commercial endeavor is based on an implicit if not explicit model, including an objective, the means to be used, and the environment within which it will be carried out. The doctor has a model of how the body functions, including what is "normal" (without which we could not define disease).

Models are the conscious preoccupation of some sciences. Astronomy and physics seek, for example, better models of how the universe "originated." The concept of a starting point is a simplifying constraint from which to start a cosmic model. It was presumably first adopted because of the difficulty of dealing with infinite history. The Bible provided the initial

---

†For an early description of some of the pioneering World War II work (included in the introduction reference to early adumbrations in the work of Archimedes, da Vinci, Vauban, and others), see the classic declassification by Philip M. Morse and George E. Kimball.[1]

conditions that governed Western thought for some two millenia—the Scopes trial took place in the twentieth century. An infinite, no-starting-point model did not seem, intuitively, to "explain" how things got started, because there was always something prior. But neither was a creative model intuitively adequate. Today, cosmology postulates an origin of the universe a finite number of years ago, but also hypothesizes that there were preceding universes.†

The concept of a model is, then, very broad and general, but always subject to constraints. Here we will set ourselves far narrower, or more stringent, constraints than do the philosophers, astronomers and physicists. The model for this casebook will be constrained, as the title states, to military applications. The cases will have a quantitative, computer-usage orientation, with an emphasis on simulations, and they will be simplified presentations of real-life models, recently or currently in use. They will be selected to illustrate a range of both problem types and modeling approaches that Air Force operational planners are likely to encounter and use in their careers.

A model is potentially useful to analysts and decision-makers because it represents the real world (or that portion of the world with which one is concerned at the moment) but does not replicate it. The latter would not help us. We already have the real world in all its glorious and frustrating complexity, and we do not wish to spend forever in futile attempts to take account of all the possible variations that the complexity makes possible, in blundering, random-walk nonsolutions, or even in intelligent but unacceptably slow experience-gathering and untested learning.

We wish, rather, to simplify particular aspects of the real world to help us solve particular problems. We don't want to represent everything in an all-purpose model that tells everybody everything, solves nothing, and takes forever doing so.

"Forever" may sound overblown, but any length of time longer than that which we have available to us, because of nature or of orders from our superiors, is effectively forever. This fact has been delightfully dramatized by Major General Jasper Welch in the phrase, "10^{30} is forever."[3] The point he made is that if a model has 100 independent variables, or inputs (a mod-

†There is a fascinating debate still raging on this point. It is now believed that there is not enough mass in the universe to make its expansion self-limiting and reversible. It is contended that nothing can be learned about the causes of the big bang now generally accepted as the origin of our universe, because the heat and compression of the first moments of the big bang destroyed all evidence concerning what went before. This is an agnostic position—we cannot know. To some scientists it becomes deistic, on the argument that, if scientific explanation reaches a dead-end, a supernatural explanation, or God, must take over. To others, not knowing does not mean that we may not eventually know, or that even if we can't there may not still be a "scientific" explanation. For a popular account see.[2]

est number by today's standards), each with only two values (e.g., 1 and 0—something works or it doesn't), then it would take 2^{100}, or about 10^{30}, runs of the model to test the sensitivity of the outcome to each possible change (or error) in an input. If computing time for each run were only one nanosecond, which is far beyond the state of the art (note that an electron can travel only one foot in a nanosecond), the total processing time would be $10^{30} \times 10^{-9}$, or 10^{21} seconds. Now, 10^{21} seconds is over 3×10^{13} years, or 1,500 times the estimated age of the universe—and surely that's forever. In fact, one or two years—less than one ten-trillionth of this time—may be forever for an analyst who has a deadline or an impatient boss. Surely it is forever if he is paying for that much computer running time.

Patently, even models that greatly simplify the real world can be very complex; we will have more to say later about the need for a "preprocessor" to limit the numbers of possibilities that are really relevant and necessary to examine.

An analogy for such a preprocessor might be the experience and "positional sense" of a chess player, which enable him to examine but a small portion of the similarly astronomical numbers of possible sequences of moves. A chess game starts off with 20 possible first moves for each player, and that number rises before the trend is ultimately reversed as the game is "simplified" by the exchange of pieces. If the theoretical number of choices remained at 20 per move, 10^{30} possible game patterns would be reached on white's 12th move. If the *average* number of choices were *only* 2.4, "forever" would be reached on black's 40th move, which may be only adjournment time in many matches. (Those who prefer the game of "go" can readily determine that by the time white places his sixth stone, the possible combinations have passed 10^{30}.) In chess, many choices are so obviously ineffective or adverse as not to require even a moment's consideration, and even grandmasters cannot foresee all the possible outcomes of even the most sensible moves. In "go", there is the additional factor that, especially in the opening phase, the outcome is indifferent to a considerable proportion of the choices because of symmetry. Similar factors limit the choices in many military models, but setting up rules for their determination can be terribly difficult, especially for computer models.

Thus, to our first constraint of limiting the scope of the models to be considered to military applications we must add a constraint on the complexity and length of time required for solutions and "runs". But reducing complexity always involves a trade-off with realism and the risk of omitting a factor that is important. This is particularly likely for factors that are not quantifiable, or not readily quantifiable. Models generally deal with numbers, even when these are handled nonparametrically, i.e., with inequalities rather than equations. But precisely for this reason, they can be, as in other contexts, dangerously seductive, playing on the analyst's natural predilection for clean-cut, satisfyingly "objective" numerical solutions.

In addition to relevance and simplicity, then, we must specify a criterion of completeness of coverage of significant factors—or of candor with ourselves and our customers about what is left out. The art of modeling is becoming increasingly sophisticated in methods for introducing nonquantifiable, judgmental factors. One such method is the introduction of a "man-in-the-loop," or man-computer interaction. Nevertheless, the analyst must be eternally vigilant against overemphasis on the numbers and must always seek to define the limits and omissions of the model as well as what the model does do to assist the analysis, and he must make the limits and omissions of his analysis clear to the decision-maker he seeks to aid.†

THE PURPOSES OF MODELING

But if models are not all-purpose and can't do everything, what can they do? We will come to the many specific kinds of problems they can attack, but first we must note that models cannot always solve problems. Indeed, many problems cannot be solved, particularly in the military field, in which, in the last analysis, answers can only be determined in war—the avoidance of which should always be a primary objective. But when they cannot provide solutions—and sometimes they can—models should always "shed light." They may do this in several ways:

The process of constructing and using a model should increase the understanding by both the analyst and his client of the process or problem being studied. It can be heuristic, helping the analyst to find ways to shed

---

†The Soviets have placed great emphasis on "mathematical formalization" and "cybernetics" in the analysis and planning of military matters, yet they also voice the above caution about the limits of quantification. The following quotations from an authoritative Soviet text are of interest:[4]

> In spite of their complexity, the process of warfare can be described by means of a rigorous mathematical edifice, with a high degree of accuracy in approximating reality.

However,

> . . . mathematics cannot substitute for an entire complex of social, economic, and ideological substantiation essential for making crucial political and strategic decisions.
>
> The adoption of mathematical methods not only does not negate the necessity of a qualitative analysis of phenomena but, on the contrary, is based on such an analysis. . . . By means of such an analysis one isolates the object of investigation, determines its composition, the character of internal and external relations, and establishes the parameters which require quantitative evaluations, and a criterion of effectiveness, by means of which one can make the necessary calculation and change appropriate characteristics of the phenomenon.

light. But the purpose is not just to educate the modeler. The learning must be transferred to the user, or decision-maker. Communication and interaction between modeler and user must be continuous and open—if light is shed, both must be able to read by it.

Models can aid in making choices—for choices must be made even when uncertainties cannot be fully resolved and solutions found. Doing nothing is choosing one alternative. Models can assist in comparing alternative weapons systems, tactics, environments, routings, training methods, and so on. Relative numbers, then, are what count in selecting among alternative means.

Models may sometimes give answers, in the sense that the absolute numbers are taken as valid. Examples, however, are not easy to come by. A limited logistics model *may* be able to give valid estimates of absolute quantities of fuel consumed, or vehicles required in given circumstances. But for the most part absolute numbers should not be believed. A bomber penetration model may give 70 percent bomber survival, or 70 percent of targets hit. We cannot know that 70 percent would be the real number if there were a war, even though we may have high confidence that the particular laydown that gave this figure will do better in a real war than an alternate laydown that gave 50 percent vice the above 70 percent when used in the same model. Models are seldom valid for force sizing, though they are often used for rationalizing decisions made on other, e.g., budgetary, grounds. In short, back to the relative numbers of the last subparagraph, above.

In the author's view, the above purposes are given in descending order of importance. One should always seek first to learn from a model. In the process of learning, one can often use a model to assist in making choices—while caution and skepticism are always in order, especially about whether the right criterion has been chosen for ordering the alternatives, there are many cases in which one can believe relative numbers. One should seldom, if ever, credit absolute results of applying models, at least in the highly uncertain world of military affairs.

The purpose of a model should always be subsidiary to the purpose of the modeler or of the decision-maker he serves. It has often been said, but is also often overlooked, that models do not analyze anything. Analysts analyze, and models can assist them in their task. Models should always come after the definition of the problem (which comes after the selection of the problem, although this may be done for the analyst). Modeling is one, but not the sole, aid to analysis. It is never clear a priori that a numerical, mathematical or computer model should be used, or that a particular type of model should be used. Selection of the means of analysis is a part of the analytical process itself.

The above may sound obvious, but nevertheless it frequently occurs that analysts apply a model they know and like when this is not the best approach to the problem. For example, much attention has been given by strategic analysts to the problem of recovery from a massive nuclear attack, especially since delaying such recovery was made a U.S. strategic objective by former Secretary of Defense Rumsfeld.[5] The most common approach to analyzing a recovery problem has been the use of econometric models, starting with input-output models and modifications thereof. But it should be clear both from historical examples (notably, World War II) and by simple reasoning that a purely economic analysis cannot tell us very much about the timing or rates of potential recovery until we know a considerable number of political and politico-military conditions. Has leadership survived and maintained or reestablished control? Has a stable currency been provided? Is the war continuing? Are other countries aiding (willingly or otherwise)? Is the enemy interdicting aid, or exacting reparations? These and many other questions must be asked prior to examining questions about the model itself, such questions as: Do the model's inter-industry coefficients take adequate account of substitution and other conservation measures, of allocation policy, and so on?

TYPES OF MODELS

As has been observed of people at the beach, models come in many sizes and shapes. And, like people, models can be categorized in many ways. In addition to the above range of purposes, four taxonomies are considered here:

Application, or subject of study.
Objective function—effectiveness, cost, and cost effectiveness.
Level, or scope.
Technique, or method used.†

Application

Interest herein has already been limited to military applications, with an emphasis on Air Force concerns (which, however, interface in important

†Many analysts would include autonomous vs. interactive, which will be subsumed here under Technique.

ways with Army and Navy forces and operations). But models may find uses in virtually every Air Force activity, at every level. The breadth of applications is suggested by the following partial listing:

Force structure
Force sizing
Procurement
Costing
Testing and Evaluation
Program management
Strategic and Tactical combat operations (weapons, systems, forces, weapon allocation, tactics . . .)
Combat support services (C^3I, mobility, medical . . .)
Support (maintenance, training, personnel, recruitment . . .)

The above list is neither mutually exclusive nor exhaustive. Some would use a Strategic/General Purpose Force/Support Functions division. Our purpose is not to argue for any particular classification, but to set up a straw man in order to search for general principles of modeling that may apply.

Though applicable, models may not be equally useful in all of the above categories. We have already commented on their limitations in assisting in force sizing, for example. A given model may also involve more than one of the above applications. There is no reason any model should not, and it may often be necessary, but the analyst will be well-advised not to try to model too many problems or serve too many users at once.

The object, or characteristic to be analyzed, may be classified in many ways. One of the most basic breakdowns is:

—Effectiveness
—Cost
—Cost-effectiveness

These broad categories may be broken down in various ways.

Effectiveness

Effectiveness is a term most often used in connection with weapon systems; it may also apply to support systems and to operations of all kinds. Its meaning varies with the level of operations, or the scope of the model, and with the figure(s) of merit (FOM) chosen for its measurement. Ideally, one wants a single FOM—targets destroyed by a missile, aircraft kills by air defense, aircraft kills in air-to-air combat, etc. Life being seldom simple, it usually turns out that no single FOM is adequate. Targets destroyed by a nuclear missile must also take account of non-target destruction, e.g., collateral killing of people—hence the now well-known "dual criterion" of some strategic analyses: DE, or damage expectancy, and CD, or collateral damage. Kills in air combat cannot ignore losses—but going to relative attrition rates may not tell us whether the kills were sufficient to meet the real objective—to prevent the enemy from carrying out the missions of his aircraft other than air superiority: interdiction, close air support, etc.

If a battle is modeled, the measure of effectiveness may be the outcome of the battle—in terms of winning or losing, or of relative attrition, or of position at the end of the battle. The outcomes of single engagements or battles may be measured in terms of damage achieved or of targets killed. Again, there may not be a single FOM that is acceptable. If we are concerned with the effectiveness of a bomber force, for example, do we measure numbers of targets hit (or value destroyed) and ignore bomber losses? In an all-out strategic attack, perhaps so, although even here we will want to consider bomber survival not only for the sake of the pilots but also because of the possibility of recycling the bombers in follow-up attacks. In conventional bombing, on the other hand, there is no question that we must consider attrition rates, for in conventional war there is always the issue of follow-on missions. But in an air superiority battle, the exchange ratio of aircraft may afford us an acceptable single criterion (if the initial forces are not too disparate, numerically). In a particular logistics model, to take another of many possible examples, we may be able to use route length, fuel consumption, or tons delivered per day as a single FOM, though multiple criteria (including, e.g., cargo size) are more likely to rear their ugly heads.

Military analysts are fond of observing how much easier industrial modeling is—the FOM is profits, commensurate with costs. But this is only a partial truth—the businessman must balance short-term versus long-term profits, and he must temper his profit motive with legal, ethical and public-image constraints. His costs are not measurable just by counting profits, since there are external costs of equal opportunity, environmental pollution, and so on. In neither the industrial nor the military case can we count on finding a legitimate weighting scheme that enables us to add or otherwise combine FOM numbers in a meaningful way, although one often sees the attempt.

Numerous measures of effectiveness may be generated automatically in simulations; the modeler may initially have all of them printed out for analysis and later program more selective printouts for different user requirements or because some FOMs have proved more useful than others.

The problem of measuring effectiveness is broader than just the question of whether one can find a single figure of merit. It can be a matter of searching for solutions to some basic philosophical questions. If we are dealing with weapons systems, for example, we may have to consider whether we can find rational ways to compare three different categories of weapons:

1) Existing weapons
2) New weapons that do what existing weapons do, but better, e.g., with
—greater accuracy
—higher kill or destruction capability
—less weather degradation
—greater reliability (fewer maintenance hours, less red-line time)
—less training time or pilot stress, etc.
3) New weapon concepts that do things that can't be done now, e.g.:
—satellites were an example two decades ago
—antisatellite weapons (ASATs)
—high-energy lasers
—directed particle beams
—VTOL aircraft
—things we haven't thought of yet.

Such disparate weapons, or weapon systems, are not easy to compare with each other. Yet choices must be made. The choice problem is exacerbated by the great difference in the time at which different systems may become available. This often means tradeoffs between better performance later and possible military risks earlier. Since there are apt also to be great cost differences, the problem is discussed further under "Cost-Effectiveness", below.

Another difficult issue is the trade-offs between quality and quantity—the F-14 and F-15 versus the F-16 and F-18, Admiral Zumwalt's "High-Low" concept of buying both some sophisticated, high-cost systems and more simple, low-cost systems, and so on. Since qualitative advantage usually involves greater cost, this topic too will be discussed further under "Cost-Effectiveness." A few things need to be said here, however. There is a straight effectiveness issue in the above example of the exchange-ratio criterion (the air-superiority-battle case). Suppose the exchange ratio is favorable to us, by, say, two-to-one, but the other side has more aircraft, by four-to-one. Clearly, our side will have to break off the battle, or lose, because in time continued successful engagements will lead to zero planes on our side and one-half the enemy force surviving. (This is almost reminiscent of the alleged words of Chairman Mao about nuclear war: "Two-hundred-million Chinese killed; four-hundred-million Chinese left!")

One must also ask whether a qualitative improvement really increases effectiveness or is simply "gold-plating". As in the case of models to which too much may be added simply because "we can do it, and someone might use it," in weapon systems some added features may even lessen effectiveness, if, for example, they make operation more difficult or maintenance (and therefore unavailability) more frequent. On the other hand, the analyst may find that some qualitative improvements have been foregone not because they are too costly but for lack of foresight or because of institutional inhibitions. For example, the Soviets but not yet the Americans have equipped many military vehicles with internal-pressure and filters, for operation in chemical and biological warfare (CBW) environments.†

Other issues, too, will have to be faced in formulating measures of effectiveness. Is a single-purpose or a multipurpose aircraft preferred? The multipurpose advocate may argue costs, but there are effectiveness issues at stake, too. What are the contingency scenarios? Is it sometimes unpredictable which purpose will be needed? Can the purposes sometimes be fulfilled sequentially by the same aircraft? Are there performance penalties in some or all of the roles, and if so, are they critical?

Similarly, the analyst may have to consider dual-service weapons or systems. The TFX argument is history. The questions of common Air Force-Navy cruise missiles and next-generation intercontinental-range ballistic missiles are current events.

Since most of the above examples—and most real-world choices—involve consideration of costs (George Bernard Shaw notwithstanding, price is never 'mere', even for the military analyst), the next section will discuss some of the principles and problems of military costing, and we will then be ready to consider the broader questions of cost-effectiveness analysis.

Cost

Cost can generally be estimated with less uncertainty than effectiveness, and costing is at least conceptually simpler than effectiveness analysis. It is nevertheless important to note that cost estimation can be a highly complex problem in itself. This complexity will be illustrated in Chapter II, but a few general observations are in order here.[6]

---

†It is interesting to note that such defense measures have been taken by the side with the offensive chemical capability, not vice versa. Clearly, (1) an offensive chemical capability requires defenses to enable operation in one's own offensive environment, and (2) the defensive capability may also reflect different evaluations of the likelihood of chemical (or tactical nuclear, or even biological) warfare. Note also that the issue is relevant for aircraft, not just ground vehicles; airbase attack with chemicals, for example, could be a profitable Soviet tactic in Europe.

The difficult problem in costing is the handling of *time*. This factor is implicit in the basic classification of the costs of military systems into R&D, Investment, and Operating costs. Research and Development (or, more explicitly, Research, Development, Test and Evaluation—RDT&E) is the first cost incurred for a weapon or other new system. It may start several years before the Investment, or Procurement, phase, although it also generally overlaps the latter, as testing plus R&D on improvements and modifications continue. For a highly complex system, some R&D may go on throughout the life of the system. R&D is conceptually a one-time investment cost. (Readers who have been in industrial organizations may be used to the "expensing" of R&D costs in the year incurred, as generally permitted in the Tax Code since the 1950s, but the cost is still of an investment type: the output is "used up" over the life of the system, and the R&D may, indeed, have considerable "salvage value"—see p. 78).

Investment, also called Initial Investment, covers the procurement of the system itself, including "initial spares." It includes items with a long life, measured in years. In a commercial enterprise, it would be "written off" by way of depreciation over the estimated life of the investment. Investment costs may also continue after full deployment is reached, as modifications of components (from continued R&D) are retrofitted to the original equipment.

Finally, operating costs—maintenance, training, personnel costs, fuel, recurring spares, etc.—are incurred during the operation of the system. Where there are many "units" of a system, operating costs start when the first unit is deployed (initial operating capability—IOC) and build up until full operating capability (FOC) is reached, when they generally level off (ignoring inflation) until the earliest units start phasing out. Thus, operating costs overlap investment and sometimes also R&D costs.

These three cost elements are sometimes called "Life-Cycle Costs," to emphasize the importance for both analytical and planning purposes of estimating the operating costs over the lifetime of the system, especially when considering possible parametric variations in levels of procurement, rate of build-up, inter alia. Strictly speaking, Life-Cycle Costing should also include estimates of salvage value. Are units of the system sold for scrap (as ships often are)? Are they downgraded to "second-line" use (perhaps sold or given to a developing country)? Are they "mothballed" for possible future use (recall the 50 World War I destroyers given to Great Britain for the Battle of the Atlantic before the United States entered World War II—and some contemporary B-52s)? Are they "cannibalized" for parts for surviving units, as airplanes sometimes are?

One of the most important salvage values is sometimes the contribution of the R&D phase (and sometimes of the manufacturing and operating experience) to the technology of future systems. This return on investment is virtually unmeasurable, but it is sometimes taken into account subjectively as a "bonus" from the selection of a particular system. One example the writer dredges up from distant memory involves the early comparison of the NIKE air defense system with alternatives such as the HAWK and TALOS (in a land-based version derived from the shipboard system). While it is impossible to say what influence it had in the NIKE development and procurement decisions, it was noted by analysts at the time that the NIKE, being a command-guidance system, would hold more promise for development into an ABM system than the semi-active and active homing systems mentioned above. It also had the advantage over the TALOS that it was rocket-propelled, not air-breathing. (TALOS was rocket boosted but ramjet sustained.) Today, terminal homing, and perhaps even air-breathing missiles, may have potential roles in ABM, but they could not in the 1950s state-of-the-art.

In theory, salvage value is always a deduction from life-cycle costs. In practice, however, it is usually omitted, for two reasons. First, salvage value is usually very hard to predict, whether one is dealing with R&D or hardware. Secondly, salvage is generally expected to be far in the future and is therefore highly discounted and in effect neglected.

This brings us to the topic of discounting, expressing time preference, often a very controversial matter but nevertheless highly important. We prefer to have goods *now.* And we properly downgrade, or discount, our future obligations as compared with our present ones because the future is less certain and because we have more time to prepare for it. Today's bill we must pay today, out of today's funds. As for future bills, we may die before they come due; if we don't, we will have more time in which to earn money to pay them, and we believe that inflation may decrease the value of our bills. In fact, most of us assume that our ability to earn money will increase over the years, and this is certainly true of governments, which do not generally retire or lose earning power with age, as individuals do.†

Hence, whenever goods, such as weapons systems, are durable and the life-cycle significantly long, and the patterns of expenditures for alternative systems differ, we should reduce the future amounts by some stated percentage, say ten percent, for each year that we project into the future. This amounts to saying that if money is worth ten percent per annum and we have to put up one dollar this year, it would pay for $1.10 worth of goods next year, $1.21 two years from now, and $2.59 ten years from now. Put

†Governments, do, of course, like individuals suffer decreased income in recessions and depressions.

another way, for one dollar we may have to spend ten years from now, we would have to put up only thirty-nine cents today, if this initial thirty-nine cents could earn ten percent a year for the next ten years, compounded. This is the standard "discounting" that economic theory indicates should be applied to all future investment decisions. Using life-cycle costs without discounting assumes a discount rate of zero and is unrealistic.†

Unfortunately, discounting is seldom used in military costing, perhaps both because it is a nuisance and because it is not widely understood. In the early 1970s, when a wave of inflation started in this country (and worldwide), Congress, like home buyers and subway buyers, became increasingly concerned with escalating costs. Congress put great pressure on the Pentagon to anticipate these costs and not to incur overruns. Without having been subject to this pressure, the Monday morning quarterback can argue that the Pentagon made a mistake in acceding to the practice of stating future costs in "then-year" dollars in addition to the "this-year" (constant) dollars that had been used in the past. Then-year (current) dollars attempt to project the future trend of inflation and to build this into the estimate. Aside from the fact that no one is very good at predicting the future rate of inflation,† this practice has the anomalous effect of making future expenditures appear more onerous than present expenditures. It actually amounts to using a *negative* discount rate, implying that we would have to put up a dollar now to get 39 cents 10 years from now (at the ten percent rate)! We would all laugh if we were offered an investment like that. But people who don't want the investment made tend to use this figure and not the cost in this-year (constant) dollars, with or without discounting.

An important example of the effect that the underestimation of inflation

---

†There is a large literature on discount theory and the practical problems of the appropriate discount rate for various classes of buyers. The literature has been reviewed, and a 10% rate found to be well-substantiated for weapons systems procurement by Shisko.[7] The arithmetic of discounting is simple, following the basic formula:

$$C' = \sum_{i=0}^{n} \frac{C_i}{(1+r)^i}$$

where C' = total discounted "life-cycle" cost
C_i = cost in ith year
r = discount rate
n = no. of years in the life of the program.

†Not only are inflation rates tough to forecast, but there is a built-in pressure for the government estimator to keep down his estimates of future inflation because he must act as though he assumes that government programs to control inflation will work. To do otherwise would be to encourage inflation. In 1976 the DoD Comptroller estimated that the inflation rate would decline to 4% in the 1980s. To the extent, of course, that the inflation-control programs do not work, the then-year dollar estimates will be too low, which is at least better than having them overestimated!

can have is seen in the U.S.-lead NATO agreement for a three percent annual increase in the *real* military expenditures of the members. The real increase in U.S. expenditures in the first year of that agreement turned out to be about one percent, rather than three, because of inflation above the original estimates. As the first draft of this volume is being revised (summer 1979), another example appears imminent. In the SALT II debate, Senator Nunn, followed by others, raised the question of increasing U.S. military expenditures four to five percent in real dollars in order to redress a perceived military imbalance before they could accept SALT II. Clearly, a four or five percent real increase, as promised, is likely to go the way of the 1978's three percent increase. A much more realistic approach, therefore, is that suggested by Henry Kissinger in his testimony, namely, that a supplemental appropriation be brought in, covering specific programs. If inflation then makes it infeasible to complete the specific programs with the appropriate monies, a future supplemental appropriation can be brought in to provide the necessary funds.

Discounting, when used, should be applied to constant-dollar estimates only. In theory, we could discount then-year dollars, using a discount rate that includes factors for both time-preference and inflation ($[1+p][1+i]-1$), where p is the time preference, or "pure" discount, and i is the expected rate of inflation. However, there are two reasons for not doing this in practice. First, projections of inflation are not reliable. Second, if projected inflation is high, p and i will interact, that is, people will tend to increase p "just in case." We will do better to base p on pre-high-inflation estimates.

Even if it were precisely predictable, inflation should not generally enter into comparisons. If inflation is the same for all commodities, it represents simply a change in all systems alike. Price changes are not always uniform, however, as everyone who buys petroleum products knows. A heavily fuel-consuming system would have become relatively less cost-effective during the 1970s than a low-fuel system which serves the same purposes. But sharp changes in relative costs are rare and certainly not highly predictable. Another exception to prove the rule has been computer costs, which for a given capability have for some years been declining relative to most other costs, so that systems with high substitution of computers for manpower have generally become more cost-effective. The Defense Department does not break down its inflation projections that finely, however. It estimates inflation for only four categories: procurement; RDT&E; military construction and family housing; and shipbuilding. For the first two, the forecasts are almost identical, leveling off at 4.0 percent a year in 1981. The last two are a bit higher, leveling off at 5.0 and 6.4 percent, respectively, in 1981.

This is consistent with history for a number of years, but one cannot be at all sure that these trends will continue indefinitely. It is not clear how these uncertain differentials can usefully affect decision-making.

Life-cycle costing runs into difficulties if the life-spans of systems differ. If the systems are truly alternative, i.e., can carry out the same missions, then discounting will not wholly solve the problem. Successor systems will have to be postulated—and costed—for the shorter-lived systems. These may be difficult to postulate and to cost, and they may outlast the longest-lived system. The process could go on and on, like the bread and the jam, becoming more and more unrealistic as it goes.

While life-cycle costing can aid in understanding the total impact of new programs, accumulating costs over considerable periods of time is often in practice used to discredit programs. With the additional loading of all related systems costs onto the given estimate, large military systems can be made to appear inordinately expensive. Before June 1977, the B-1—projected over a thirty-year life, with inflation, and loaded with all the weapons it could carry and the tankers that would service it—was called the "most expensive" weapon system in history. Then the Trident boats and missiles vied for this honor against the MX missiles in tunnels. At the time of writing, the title is being bestowed on the nuclear aircraft carrier. This is not to say that these are not expensive systems, nor to enter into the controversy as to whether they are too expensive or indispensable, but simply to point out that the analyst has an obligation always to be precise about the meaning of his cost estimates, to use discounting where necessary (not the projection of inflation), and to make comparisons on the basis of relative *incremental cost,* not the total cost of systems expanded to include many common elements or items that will be purchased regardless of the decision.

We shall return in a moment to the definition of incremental costs. First, a further word on the handling of time. Discounting is a first step in measuring streams of costs over time. The problem is minimized if comparative alternatives are available simultaneously. When they are not, the problem is one of *availability.* Too often, the problem of availability is ignored by the analyst, although the decisionmaker must decide whether he can accept the risk of waiting for "something better". "The boss wants a number," and analysts tend to give him one. The Air Force used "five-year system cost" (R&D plus investment plus five years of operation) for many years. The concept was developed at Rand in the early 1950s, on the assumptions that most aircraft had a five-year "first line" life (hardly descriptive of the B-52) and that the great simplification of this formula would facilitate both the work of costing and the understanding thereof. More prevalent today is the ten-year system cost (R&D, plus investment, plus ten years of operation), which dates more or less from the 1965 DoD efforts at standardization among the three Services, in connection with the "Tri-Service Study" of ballistic missile defense. Many years ago, the writer advocated the more explicit treatment of time by the presentation of annual costs for the years in which they are expected to be incurred. In cost-effectiveness studies, this presentation must, of course, be accompanied by estimates of effectiveness over real time.[8]

Note that we speak of "expected to be incurred". It is axiomatic that only future costs should affect decision-making. Costs incurred or committed before the decision date are "sunk" and should never affect decisions. If past investments make large increments in effectiveness possible with small future expenditures, this will automatically show up in the analysis, but it should not be assumed on a priori grounds. Since the decision date is presumably after the analysis, the analyst should make explicit the data assumed in the modeling.

The analyst must take special care to insure that costs committed on the decision date, but to be expended later, are treated as sunk. For example, when the fifth and sixth B-1s were cancelled it was widely alleged that roughly one-and-a-quarter billion dollars would be saved (and that each of the aircraft would therefore have cost over one-half billion!). In point of fact, cancellation costs and some approved R&D costs amounted to some three-quarter billion dollars. This was sunk cost, and only about one-half billion of future cost was saved. (And the cost was still not properly called one-quarter billion dollars "per aircraft"—part of the cost would have been for further R&D and testing, and for keeping future options open.)

But to return to the question of incremental costing. If costing is done for the purpose of comparing alternatives (including the alternative of buying nothing), then what count are the costs associated with buying A, or nothing, or B instead of A, etc. These are the incremental costs of making a given decision. Elements common to A and B should not enter into the comparison of A and B. (If there are elements common to A and no new system, then clearly these are items that would be purchased regardless of the decision.) If two aircraft will deliver the same missiles or bombs, then the missiles or bombs do not need to be costed—unless A and B will deliver different total numbers of missiles or bombs.

People—and analysts are people—are often reluctant to cost incrementally only. They like to anticipate the question, "What will the *total* cost be?" If the only purpose is the comparison of alternatives, then this is not a relevant question. In fact, the higher the total relative to the difference, the more total costs obscure the advantage of one alternative over another.

There may be many legitimate reasons for needing total costs, but the analyst should make sure they are relevant to the question he is analyzing. Otherwise, he will find his work used, or misused, for purposes other than those intended. Again, the analyst must always be on guard against attempts to make the model serve too many purposes or too many users.

Reliable cost data are seldom easy to come by, but they are still, in general, easier to obtain than effectiveness data. The fact that this is well known may indeed introduce a problem for the analyst. He may have to be sure that the costs, particularly of off-the-shelf items, are given with a degree of precision not warranted in relation to the range of uncertainties of many other elements of cost, not to mention effectiveness, in order to avoid being discredited for insignificant inaccuracies. Nothing is so disconcerting as receiving a memo—or a challenge at a briefing—to the effect that "you have such-and-such a truck (radio, radar, etc.) in there at $8000, when I know for a fact that in the last Air Force order, they cost $9246.73 each." The analyst is compelled, therefore, to strive for great detail and precision in order to have a protective verisimilitude.

In general, the required data will come from two sources: (1) The Services, for the above-mentioned truck price and, more importantly, manpower and operations costs; (2) industrial suppliers, to whom we must turn for help in estimating such items as R&D hardware, contractor testing, and procurement. There will always, of course, be the suspicion of self-serving inaccuracies, e.g., bidding too low in order to obtain a contract. One of the most difficult problems for the analyst is, therefore, in the case of future procurement to allow adequately for cost escalation from contractor increases and government-imposed design changes. (Inflation per se has been covered in the above discussion.) Wherever possible, the analyst should endeavor to validate past estimates by comparing them with actual contract costs as programs progress.

Note that we have been speaking of peacetime costs. These will be the principal concern of the analyst. But it should be recognized that in wartime, values change. In limited wars, or in long all-out wars, new weapons and other systems may be brought in, and the above principles should still apply to making choices. But at any given point in a war, the choices are those of a commander, not a planner. These choices involve alternative deployments and employments of physical resources or assets—not dollars—on hand. Costs are the opportunity costs of not being able to use given assets for other purposes, once they are committed. Peacetime costs are opportunity costs, too, but they are the sacrifice of the best alternative purchase with the given amount of dollars.

In terms of analytic models, then, the wartime analogy is that of an effectiveness model without costs. Given so many aircraft of a given type, what is the most effective way to utilize them? This may be, for example, a matter of optimizing the allocation of weapons to targets, or to committing fighters to air combat, or of allocating multipurpose aircraft to different roles.

Cost-Effectiveness

The term cost-effectiveness was first used in post-WW II military operations research, in recognition of the fact of life that military effectiveness cannot be measured in dollars and so it is necessary to measure the effectiveness that can be achieved for a given cost, or the cost of achieving a given level of effectiveness. It is interesting to note that when the Program Planning and Budgeting System (PPBS) was mandated by President Johnson for all Government agencies (in addition to the Department of Defense, where it originated) cost-effectiveness studies, the backbone of PPBS, came to be known as cost-benefit studies. The latter term reflected the belief that in most nonmilitary cases effectiveness could be assessed in terms of dollars, i.e., in benefits commensurate with costs. In practice, this was not as easy as it sounded, as "external costs" such as pollution could not always be measured in dollars.

It is not possible, conceptually, mathematically or practically, simultaneously to maximize one variable, effectiveness (or benefit), and minimize another, cost. Hence, it is always necessary to fix one and vary the other. Because fixing cost and fixing effectiveness are conceptually the same and mathematically dual, it is often thought that the choice is one of indifference for the modeler, and, in fact, both approaches are used. In practice, however, the writer believes there should be a strong preference for fixing cost, for two reasons.

First, it is generally possible to define costs with reasonable precision and in terms of a single FOM, dollars, although at times other costs, such as numbers of men, may be required. Effectiveness, in contrast, can seldom be so clearly defined. Even if a single figure of merit can be found, effectiveness may vary widely in different scenarios, for all of which a single force or weapons system may have to be prepared. And different scenarios for the same forces may even require FOMs. This is an argument, therefore, for fixing cost (dollars or men in peacetime and available resources in wartime) and then varying effectiveness. One can then vary the scenario and show how effectiveness may vary, allowing the decision-maker, with whatever help the analyst can give him, to assign probabilities or other measures to alternative scenarios. It is often noted, for example, that all-out "out-of-the-blue" nuclear surprise attack is one of the least plausible of war scenarios, but that its damage would be so horrendous that it must be prepared for, even at the expense of protection against lesser but more likely attacks.†

Second, though they are complex, costs are better understood and can be better modeled than effectiveness, with less uncertainty, or narrower confidence bands, in the results. This argues further for fixing costs and warning the user of the uncertainties in the effectiveness estimates, meanwhile striving to improve the modeling and the estimates, and striving to insure that relative outcomes are less uncertain than absolute levels.

In any event, it is important to note the limitations of cost-effectiveness models. While they can teach us a great deal—the first purpose stated in the earlier section on that issue—they seldom offer a good basis for choices among, say, weapon systems unless the differences are very great. How great they need be is in part a function of distance into the future that we are projecting. For comparing existing weapons against existing enemy defenses, for example, fifty percent or less may be a significant difference. For new systems still in R&D, against defenses that can change both qualitatively and quantitatively while the new systems are being developed and deployed, an order of magnitude difference in cost-effectiveness may be required. However, preferences among systems for which the C-E ratios vary this greatly are likely to be analytically or intuitively evident without an elaborate cost-effectiveness model.

Two examples of the above may be cited. In the early 1950s, when it first became apparent that usable surface-to-air missiles (SAMs) would shortly become operational, extensive air defense studies were undertaken, notably at the Rand Corporation and the Operations Research Office (ORO) of Johns Hopkins University. These did indeed show order-of-magnitude, and greater, advantages in SAM cost-effectiveness over that of antiaircraft guns. This range justified the go-ahead on procurement and deployment of the first systems available, and the differences among systems were even sufficient to justify some choices for development of second-generation systems. Even so, there were serious oversights in the analyses. Developments in low-altitude attack capabilities were underestimated and the potential synergism between guns and SAMs was not perceived. (This synergism was apparently noted by the Soviets earlier than by the Americans—whether from better modeling or because of the Soviet propensity to retain old systems while deploying new ones.)

---

†Whether this preparation should be by deterrence-only, or should include damage-limitation should deterrence fail, is a separate argument.

Our second example is a 1977 study which showed that air-launched cruise missiles (ALCMs) would be 30 to 40 percent cheaper for equal effectiveness than the B-1 with short-range attack missiles (SRAMs) and gravity bombs. Based partly on the results of this study, procurement of the B-1 was cancelled.[9] The writer urged in 1976[10] that such a C-E range was not adequate to provide a basis for choice between systems to be deployed in the 1980s. The uncertainties about the penetrativity of the new cruise missiles against defenses that would have time to adapt to them, and hence the numbers of weapons that would be required to achieve the alleged equal effectiveness, were too great to permit conclusions of such precision. Moreover, the costs of the ALCM, a new weapon still in development, and the costs of significantly modifying existing aircraft to serve as platforms or of developing and building a new cruise missile carrier were far more uncertain than earlier studies implied. Nevertheless, these costs were compared to the costs of the B-1, three of which had already been extensively flight tested, without any allowance for potentially-required system growth in the new ALCMs (which now are being extensively redesigned). The issues of the designs, costs, availability schedules, and effectiveness (survivability of the carriers and penetrativity of the cruise missiles) are still unresolved at time of writing (late 1980).

There is another lesson for analysts in the B-1/cruise missile case, namely, that the most painstaking analysis does not always carry the day. The 1977 study was conducted by a special DoD committee. Its calculations superseded the vast amount of earlier work discussed in Chapter IV and were used as the rationale for the decision to cancel procurement of the B-1.

The above examples again bring us to one of the most common problems in cost-effectiveness comparisons, namely, *availability.* If a new weapon system is physically feasible ("by the laws of physics"), its availability is generally a function of time (and often of conditions of urgency and level of effort). The issue confronting the analyst is that very often comparisons and choices are made among systems available at different points in time, in terms of both initial operational capability (IOC) and full operational capability (FOC). From an analytical point of view, cost-effectiveness models tend to conceal a key part of the problem of choice that can only be resolved subjectively by the decision-maker: if a new system, say, System B, will be available later than System A (which may already exist or be in a more advanced stage of development), but System B has a higher effectiveness and a higher E/C ratio, then the question is, should the decision-maker wait (to develop, procure, or deploy) until System B is available? The answer is a function of the threat before System B becomes available and of the decision-maker's estimate of the likelihood that the threat will move from latent to overt status in that period. Overt, in this usage, may mean either actual warfare or crisis; in the latter case the military balance may affect outcomes in terms of deterrence or diplomatic coercion.

There is no fully satisfactory solution to this problem of differences in availability dates. Cost-effectiveness models are by their nature basically static. Costs, as discussed above, can be estimated over time and expressed in terms of annual outlays or present values. The effectiveness of weapon (or support) systems can seldom be estimated over time with comparable precision, but clearly effectiveness is zero in years in which the system is not available. At a minimum, the analyst has an obligation to indicate very clearly to the users of his analysis the variation in predicted availability dates of the systems compared. In addition, as noted above for costs, there will be widening confidence intervals for the estimates of effectiveness, as availability is further in the future. (The uncertainty of estimation of performance and costs will be compounded by the increasing uncertainty about the enemy threat, including potential countermeasures, as the analysis moves further into the future.)[8]

The problems of selecting FOMs, discussed above under Effectiveness, may seem superficially to be simplified when C-E ratios can be used. In point of fact, however, not only do all the measurement problems connected with E (in the numerator or denominator) persist, but also the problem may become even more difficult to understand. Dollars per kill (of an aircraft, a tank, or other target), or kills per dollar, may be simple and appealing measures, but they may also obscure other crucial factors. First and foremost is the matter of scale, mentioned earlier in the discussion of kill ratios as a measure of effectiveness. We have noted that efficient kills avail little if the enemy outlasts you.† C-E ratios—especially optimistic ones—should not be permitted to bias force-level planning.

The difficulty was also seen in the above example of the 1950s analyses of surface-to-air missiles that showed orders-of-magnitude lower dollars per kill for SAMs than for AA guns, but obscured the fact that SAMs could not yet cover low-altitude threats.

It has facetiously been suggested that on present trends we will be spending the entire GNP on one airplane in something like another century. It will be a plane, we assume, that can defeat any other plane (or SAM) in the world—but it can still be in only one place at any one moment and so cannot defeat *all* the other planes in the world. But this reductio ad absurdum

†The Army has tried measuring ammunition expended per casualty. In numerous WW II battles the German ratios were lower, but they lost the battles as well as the war.

argument does not tell us how far to go or what mix to buy. That is a major challenge to cost-effectiveness modelers as well as decision-makers. What is "best"? How does one tell? When must "fixed cost" be increased? There are no easy answers, but the questions are susceptible to analysis.

This basic issue keeps recurring: how far can quality go in offsetting quantity? Can American technology offset a higher level of Soviet mobilization? Is it correct to push the state-of-the-art in weapons and try to have the best airplane, the best tank, the best missile that money can buy? The issue is raised at several points herein, not to suggest answers but to emphasize its importance. It will be a dominant question for analysts to wrestle with in the foreseeable future.

Level or Scope

The scope of a given model may be expressed as the level of optimization attempted. In many cases this is dictated by the level of responsibility of the analyst, or of his superior officer, or of the unit to which he is attached. More likely, it is a matter for decision in the course of formulating the problem. In logistics, for example, the choices may run all the way from modeling the supply of fuel to one squadron or to one aircraft on a given mission, to the overwhelmingly large problem of supplying a war in Europe or of supplying the whole of the Armed Services in peace and war.

Air combat models offer us a useful "model" of a way to categorize models in ascending order of complexity and of potential learning about the processes involved:

One-on-one

One-on-one encounters are the guts of the air combat problem. Their outcomes depend on the probabilities of acquisition, maneuver into favorable position, lock-on, firing first, etc.—adding up to the probability of kill by one or the other aircraft and reflecting both chance and what is known of the characteristics of the pilot, the aircraft, and the air-to-air missiles or other armaments involved on both sides. One-on-one models may pit similar, or comparable, aircraft against each other, or one side may have a superior system and thus win most of the engagements. But one-on-one may not be realistic in an air battle.

Few-on-one

The aircraft in combat may find one or more added enemy aircraft "on its tail". What are the chances for his escape? For achieving one or more kills in the process? Or, if the single aircraft is superior to each of the several in opposition, how long does he accumulate one-on-one kills? Are there real world limits on how long a pilot would actually fight? His superior characteristics might make excessive kills an artifact of the model, without the introduction of a realistic engagement time constraint.

Few-on-few

Here, realism may increase. Our lone hero has buddies coming to the rescue. Now that we must be concerned with target assignment on each side, how do the subsumed one-on-one and few-on-one engagements add up? How do the two groups break off, or disengage? Few-on-few may also involve interactions of dissimilar systems—potentially, more than one type of aircraft on each side and ground air defense systems (SAMs and AA guns on *one* side.

Many-on-many

Many-on-many reflects a further step in realism—and complexity—which must include all of the above, with rules for combining these individual and group engagements. Many-on-many is also qualitatively different, in that command decisions (whether by actual command or doctrine, or both) must be modeled. Incidentally, sometimes the command decision model is inherent in the mathematics used and may be unrecognized by the analyst. The use of the Lanchester square law,[1] implicitly assumes that the survivor of a one-on-one encounter can be immediately redirected to engage another enemy aircraft. If kill probabilities per aircraft are low, the outcome is not highly sensitive to target allocation. Thus, if there are ten aircraft on each side, and the probability of a Blue aircraft killing a Red aircraft, $P_{rb} = 0.1$, then the expected number of kills is 0.65 if all ten Blue aircraft attack the Red lead aircraft [$1 - (1 - 0.1)^{10} = 0.65$]. If there is perfect command control (say, from an Airborne Warning and Control (AWACS) plane), and each Blue plane attacks a different Red plane, the expected value is 1.0 kill, the sum of ten independent expected values of 0.1. However, if $P = 0.9$, then ten-on-one will give an almost certain kill of one, and only one, aircraft, but ten one-on-one engagements will give nine expected kills. (Note that we have ignored the shooting oneself in the foot problem—the finite probability that a Blue plane will bag a Blue plane, if ten Blues attack one Red nearly simultaneously, since we have considered kills of Red aircraft only, not an exchange ratio. The case might have been more realistic for ten Red SAMs on ten Blue planes, with no Blue ability to attack the SAMs.) If, in the plane-on-plane case, $P_{br} = P_{rb}$, then perfect allocation of one-on-one gives a standoff, regardless of the value of P. But if Blue can succeed in massing his forces, so that ten Blues attack one Red without being attacked by any of the other nine, then Blue will start to win a battle of attrition; if P is low, the probability of the loss of one aircraft will still be high for Red and low for Blue.

Theater

At the theater level, we may no longer be satisfied to model who wins each engagement. We need to know how outcomes affect the theater war—interaction with the ground forces, movement of the FEBA, and other measures of progress toward victory or defeat. Theater models are the most challenging of all, and no fully satisfactory theater model has yet been constructed. As will be seen in Chapter VI, theater models cannot usefully predict outcomes, but they can show the sensitivity of outcomes to changes in forces and policies.

Hierarchical

Because the complexity of a theater model is beyond the direct conceptualization and manipulation capabilities of most of us, and of our computers, one is driven rapidly to the concept of a hierarchy of models, in which a theater or global model makes use of a pyramid of lower level models of individual and group battles making up the war. Needless to say, great care must be taken to ensure that at each step the models are designed to be compatible, in the sense that the next higher level of the model can accept as input the output of the lower level models and there are no contradictory or inconsistent assumptions.

Technique

The techniques, or methods, of modeling are legion, and their numbers are sophistication constantly growing. There are many ways in which techniques can be grouped. A useful classification was suggested recently by L. J. Low:[11]

Analytic games
Computer games
Interactive computer games
Computer-assisted war games
Manual war games
Military exercises

Low limited his remarks to "games," but still he might have extended his spectrum at both ends.† At the low end, he might have added purely analytical, closed-form solutions, as in the Lanchester attrition equations. At the upper end, he might have added "small wars", in which nations try out new weapons and tactics. Classic cases are the "peninsular wars", on the Iberian Peninsula in the late 1930s and the Sinai in 1967 and 1973, and to some extent also in Vietnam. One might also cite the small European wars of the mid-19th century, in which new rifle developments figured conspicuously. Needless to say, such "gaming" does not seem appropriate for testing nuclear weapons and doctrine.

As Low pointed out, operational realism and human decision impact rise as we ascend the above gaming ladder (reading *down* the page, above) at the expense of decreasing degrees of abstraction, outcome reproducibility, and convenience and accessibility (i.e., increasing cost). Our cases will fall largely in the middle, in computer simulations (usually games).

Also not separately noted in the above classification is the possibility of using the rapidly-developing programmable hand calculators as a substitute for and supplement to computers. The analyst or planner assigned to an "analytical shop" is quite likely to want one of these little marvels as an aid in his own work, experimentation, and learning. While the machines have become remarkably "cheap" for what they can do, the good and flexible "HPs" and "TIs" still cost in the hundreds of dollars and may represent a trade-off with a color TV for the individual who must buy his own. We will not attempt to include hand calculator examples in the following cases, but we call the reader's attention to a recent Rand report on the subject that includes a considerable number of HP-67 programs of military relevance, cites a Personal Programmers Club of over 2,500 members, and invites the submission of programs, inquiries and suggestions to the author of the report.[12]

Patently, this classification by technique can be crossed with the earlier classifications in a multi-dimensional taxonomy of models. Not all intersections, or cells, would be of equal interest. For example, one would be unlikely, and generally unwise, to base force planning on low-level (e.g., one-on-one) analytical models. On the other hand, a theater level or global

---

†In a footnote,[11] Low said: "The writer recognizes that many analyses are concerned with 'suboptimization' problems involving man-machine system performance in military operations environments where the existence of an adversary is only implied. Whenever the existence of an adversary does not enter explicitly into the definition of the problem, the solution technique would not, for purposes of this discussion, qualify as a game." It is useful for our purposes, however, not to exclude games in which the adversaries are not two countries but competing objectives, e.g., aircraft speed versus maneuverability, delivery time versus cost, and so on. We also do not exclude autonomous models that lack the interactive properties of games.

model is unlikely to be used to solve a problem of personnel training or fuel consumption.

These several ways of classifying models have been offered simply as an introductory framework. The case chapters will illustrate aspects of each classification, but not necessarily all categories or possible combinations.

APPROACH

The approach to case studies in the following chapters will parallel what the writer believes is a sound general approach to military analysis and modeling, to the extent that information is available on each step in the examples used. It must be noted, however, that complete documentation of all the steps is seldom available. Significant modeling efforts generally involve a number of people over a considerable period of time. They are therefore costly, and few organizations care to add the cost of historical write-ups to the cost of getting the job done. Moreover, organizations do not necessarily cherish permanent portraits that might show all the warts. Modeling efforts usually interact with institutional interests and are therefore unlikely to be blemish-free even if the analysts make no technical errors. If there have been institutional or political constraints on the analysis, involving the protection of institutional sacred cows or avoiding those of others, it may be very difficult to find a written—or oral—record thereof. It has been said that sacred cows make the best hamburger, but it may be very hard to get at the meat thereof. Moreover, there is a tendency to believe that models made by a particular group or institution, whether one of the Services or otherwise, are black boxes hiding clever schemes to protect the biases of the modeler. Even if a good and honest job has been done in avoiding biases, the proprietors of models tend to be understandably defensive on this point, and objective history may still be difficult to elicit.

Formulation of the Problem

It may seem trite to say that analysis starts with a problem needing a solution. But that is a very simplistic statement of a very complex matter. The problem may be assigned—because the boss was overheard to express some interest in it, or because of a Congressional inquiry, or because it has been "obvious to everyone" for some time, or for any one of a number of other reasons. The problem may, on the other hand, be self-generated by the analyst or analytical group, because it arises out of previous work, a need is perceived, a question is "interesting", etc.

In any case, the important point is that the first statement of a problem is often not correct, or not the best way to look at matters. Certainly, it should *never* be assumed that the initial problem, statement or question is correct. On the contrary, the first and often one of the most important tasks of the analyst is to examine the formulation of the problem and the consequent research questions generated. By redefining a problem, the analyst may make a contribution even before he goes to the trouble of designing and exercising a model. All that this step requires is *careful* study, *full* understanding of the organization, the mission and the equipment or other "material conditions," *great* intelligence and imagination, and *saintly* tact (since bosses seldom like to be told they ask the wrong questions). We assume all of these qualities in the humble analyst.

A classic example from industrial OR—with a clear parallel, but opposite drives, in military logistics—is the common management question, how can we reduce inventories? The company treasurer or comptroller can always be counted on to explain the high cost of money and the importance of reducing interest charges on capital tied up in inventory. The analyst, however, will always find it useful to respond to this initial question by asking himself, how can we *optimize* our inventory policy? Many inventory studies have shown that if proper account is taken of the cost of "outages"—failure or delay in servicing customers, cost of fabricating certain items, etc.—it will be more profitable (i.e., optimal) to increase, not reduce, inventories. The tactful analyst, of course, keeps the rephrasing of the management question to himself until he has convincing results.

Development of the Model

Making the heroic assumption that the problem has been properly stated (and recognizing that it may get restated as the study progresses), we get down to the nitty-gritty of modeling the process to permit the derivation of solutions. There are no simple rules for this task.

If we once more make an heroic assumption, that one or more FOMs have been properly selected, a technique must be found and a model developed that can yield valid, unbiased estimates of FOM values.

At the risk of being trite, it must be noted that unbiased FOM estimates assume both good modeling and collectable, valid data. Without good data, the only model available is GIGI—"garbage in, garbage out", also called "garbage in, gospel out." But the analyst will always be constrained by the costs of data collection, of model development, of inputting the data, and of running the model (no 10^{30}-nanosecond runs!).

Models do not come easily. "Canned" models have limited application, and should not be selected just because they are available: the problem, not the tool, comes first. The interesting cases of model development are learning processes, and they may take many man-years. They tend to evolve, and the analyst must be eternally vigilant that additional demands are not imposed by different users as he goes along, for these demands will often impair the performance of the model in its principal mission, whether by making the runs too long or by introducing complexities that obscure understanding of the process and the results. In short, general-purpose models should generally be avoided.

If models must usually be tailored to the problem, or questions asked, it is also true that they interact with those questions. They may answer some questions—or requirements—while modifying others and generating new ones, much as weapons systems may generate new requirements. Models tend, therefore, to be evolutionary. They may, like weapons systems, have a life-cycle, though the average length of this cycle, or of cycles for classes of models, is not yet clear.

Certainly, models—at least, large, complex ones—tend to have "mods." These modifications may—again, as for weapons systems—be driven in part by a technological imperative. The rapid growth in the capabilities of computer hardware and software presents great temptations, and sometimes opportunities, to analysts. So, too, does the continued progress in applied mathematics and modeling methodology.

In the case histories that follow, it will be seen that large models may follow varied courses. There are no agreed rules for when a model is sufficiently modified to become a new "Mark," for when the name of a model is changed, or for how much the problem can be made more manageable by modularization. It is much easier to add than to remove a mod, or component, from a simulation, which fact abets the tendency of models to grow.

But certain elements appear to be common. Models, unless aborted early, do tend to be evolutionary, to grow, to acquire a life of their own. If they do not die from competition, they often require maintenance even when not being significantly modified, and their data bases always require maintenance.

Developing and maintaining models is costly. It can be tremendously instructive and informative, hence profitable. It also carries risks—of being misleading, of misinforming, or simply of spewing forth more output than analysts can assimilate and users profit from.

Credibility

The neatest analysis in the world is of little use if it is not convincing. We must now add clarity of logic and exposition to the qualities of our humble analyst.

We have tried to anticipate this stage in the earlier injuction not to serve too many masters. If the analyst yields to the pressures to serve too many masters and too many needs, at worst he will serve none of them and at best he will have all of them to sell his product to, undoubtedly with a different sales pitch required for each.

Having a salable product *may* be insurance against having the decision made for political reasons rather than on the basis of hard findings. If our humble analyst bets on this too often, however, he may also be a broke analyst; but he must make every effort to improve the odds.

Ideally, every model should be validated. Validation is not easy and may be very—or too—costly. It has been suggested that validation should be undertaken at three levels.[13] First is the "in-perspective" aspect: Does the model address the appropriate questions? Does it go to the appropriate level of detail? How does it compare to alternative models for a given purpose? Second, "in-principle" review asks if the theoretical underpinning is adequate, the simplifications and assumptions a priori usefully realistic. And third, the "in-practice" stage examines the model's performance in dealing with actual problems and tests its predictive capabilities against observed data—which, as we have observed, may be difficult or impossible in many military cases.

When all is said and done, the analyst may or may not succeed in validating the model to the satisfaction of potential users, or in selling the use of the model's conclusions, however well-validated. In the 1960s, the NIKE-X ballistic missile defense system was one of the most extensively modeled and analyzed weapons systems in history. One could also say that the models had been quite well-validated up to, but not including, the in-practice phase. Nevertheless, in 1967, when Secretary of Defense McNamara was politically compelled to "let the nose of the Army camel under the tent," he approved none of the options suggested by model results, but chose the "thin area defense" against China, that is, against a remote future threat rather than against the more plausible possibilities of accidental launch or light Soviet probe. He added that the system might be extended to the defense of ICBM fields, a mission for which the NIKE-X system was not a cost-effective design. When a new administration came into office in 1969, Defense Department analysts used "back-of-the-envelope" calculations to demonstrate the need for the latter, and the light area defense was dropped in favor of missile-site defense under the new system name of Safeguard. The real engine of change appeared to be not the back-of-an-envelope analysis that was used as the rationale, but a political need to have a changed program bearing the stamp of the new Administration.

CONCLUSIONS

Finally, the problem has been formulated, a model created, data collected and inputted, the model exercised, sensitivity tests run, and validation attempted. It will not all have been so neat, to be sure—there will have been learning and feedback, an evolving model, and perhaps also problem redefinition. But, it comes time to ask, what has been accomplished?

First, the analyst(s) should have learned a great deal. He should have a much better understanding of the process he has been modeling. The education of analysts is only an intermediate objective, however. We must therefore ask, specifically:

—Has the learning helped to solve THE PROBLEM?

—Has the learning proved transmissible to decision-makers?

—Does the model produce credible findings? (Absolute numbers? Relative numbers?)

—If credible, have the findings been sold and used?

—Has the exercise revealed gaps in knowledge and understanding?

—Has the experience defined next steps (in problem formulation, in model design, in data requirements)?

In the cases that follow, we will discuss some real-world models, most of which have been, by the above criteria, relatively successful. The examples will be presented in ascending order of complexity and difficulty (in the author's judgment), and each will illustrate different analytical problems.

The first case is a project currently under way to construct a family of life cycle cost models and shows some of the trade-offs that must always be made, not just in costs but also in cost models.

The next case is the history of a protracted attempt to model air- and sea-lift. It illustrates both the complexities of modeling logistical problems and the tendency of many models to grow and evolve (with experience, user demand, and computer technology).

Case III describes an approach to the problem of strategic bomber penetration. The model is vast in its detail and complexity (although those tasked with tactical air battle modeling will tell you that strategic problems are simple and that analysts like to work on them for the same reason that drunks like to look for lost car keys under street lights). The model is both deterministic and probabilistic. It has been used to compare major system alternatives (but choices may still be dominated by political decisions) and for a large number of more limited analyses, in which it is regarded as a powerful learning tool.

Chapter V takes up the "tac air," or tactical aircraft, problem. The alternative mixes and uses of aircraft are studied at the level of a theater air battle, but it is not a theater battle model, since the ground battle is not modeled, only the attrition of ground vehicles and air defenses.

The full theater-level warfare problem is attacked in Chapter IV. The tensions between representations of the roles of air and ground weapons, between useful detail and necessary aggregation, between past empirical data and future wars, between desired answers and feasible learning, are dramatically revealed.

Nuclear war is the subject of Chapter VII, on nuclear exchange models. Three models are examined and it is concluded that none are useful in evaluating outcomes of potential nuclear wars, because (1) there are unsolved data-base problems and (2), more fundamentally, the problem has not been adequately conceptualized.

The last case is that of recovery from nuclear attack, Again, different approaches are considered. All are rejected because (1) all start from fairly adequate prewar data bases but none is able adequately to measure post-attack changes in economic behavior with respect to the changed data bases reflecting surviving assets and (2), again more fundamentally, the problem has been misstated—the survival and reconstitution phases, which are preconditions for the initiation of economic recovery, have been ignored or are assumed away.

In the final Chapter IX, an attempt is made to suggest a few things that may be learned from reviewing this limited selection (not sample) of modeling cases, and some of the directions that future modeling efforts may take.

References

1. Philip M. Morse and George E. Kimball, *Methods of Operations Research*, OEG Report No. 54 (Washington, D.C.: Operations Evaluation Group, Office of the Chief of Naval Operations, 1946).
2. Robert Jastrow, "Have Astronomers Found God?", *New York Times Magazine*, June 28, 1978.
3. Maj. Gen. Jasper A. Welch, Jr., "Some Random Thoughts on Models and Force Planning," a talk given at the Theater-Level Gaming and Analysis Workshop for Military Force Planning (Leesburg, Va.: Xerox International Center, 27 September 1977).
4. *The Philosophical Heritage of V. I. Lenin and Problems of Contemporary War*, General-Major A. S. Milovidov, editor-in-chief, Air Force translation (Washington, D.C.: U.S. Government Printing Office, No. 0870-00343), p. 280, p. 235. Published in Moscow in 1972, awarded the Frunze Prize in 1973, listed as recommended reading in "The Soldier's Bookshelf" section of the Soldier's Calendar for 1974, and still referenced as a most important text as recently as 1978.
5. *Department of Defense Annual Report, Fiscal Year 1978*, Donald H. Rumsfeld, Secretary of Defense.
6. Gene H. Fisher, *Cost Considerations in Systems Analysis*, American Elsevier Publishing Company, Inc., New York, 1971. This report, prepared as a textbook or training manual

for defense analysts, is a comprehensive and useful text emphasizing the complexity and uncertainty of military costing.
7. Robert Shisko, *Choosing the Discount Rate for Defense Decision-Making*, R-1953 (Santa Monica, California: The Rand Corporation, July, 1976).
8. Richard B. Foster and Francis P. Hoeber, "Cost-Effectiveness Analysis for Strategic Decisions," *Journal of the Operations Research Society of America* (November 1955).
9. Press Conference of Secretary of Defense Harold Brown, July 1, 1977.
10. Francis P. Hoeber, *Slow to Take Offense: Bombers, Cruise Missiles, and Prudent Deterrence* (Center for Strategic and International Studies, 1977), Second Edition, 1980.
11. L. J. Low, "Theater-Level Gaming and Analysis Workshop for Military Force Planning (Concept and Plan)," (Stanford Research Institute, May 1977).
12. Edwin W. Paxson, *Hand Calculator Programs for Staff Officers*, R-2280-RC (Santa Monica: The Rand Corporation, April 1978).
13. Robert E. Pugh, *Evaluation of Policy Simulation Models: A Conceptual Approach and Case Study* (Washington, D.C.: Informational Resources Press, 1977).

CHAPTER II

Costing: Life Cycle Cost Models

INTRODUCTION

We have chosen to do our first case study in the area of costing for the basic reason—at the risk of offending cost analysts, some of whom are our best friends—that it provides one of the simplest modeling problems with which to begin the learning process. Costing can indeed be complex and burdensome, and can require multiple skills for modeling, data collection, and estimation. Nevertheless, the problems are conceptually far simpler than those of measuring effectiveness, estimates are subject to far smaller errors (regardless of the screams of Congressmen and others shocked by cases of cost escalation), and the estimates are ultimately more fully verifiable.

There are, of course, many reasons why costing is an important topic in military applications of modeling. Costing is essential for management purposes, since the military services must live within budgets which they, even as thee and me, generally find too low and constricting. Costing is also essential to the making of cost-effectiveness models as a vital input to many military choices. There has also been a growing concern about costs in the past decade-and-a-half, for two reasons. First, there was the Viet Nam war, the costs of which were onerous and were borne in the first instance by cuts in other parts of the military budget, notably, the Strategic Forces, before "guns-versus-butter" choices were made via increased taxes. The buildup of the Viet Nam war also coincided with the new social programs of the "Great Society" and, as the social and political reactions to Viet Nam set in and American participation in the war wound down, the "shifting of priorities" from military to social budgets become dramatic. At the same time, the partially-related phenomenon of accelerated inflation contributed to cost-over runs and increased cost conciousness (a phenomenon not confined to the military).

LIFE CYCLE COSTING

For the above reasons, and others which we will come to, the case study used will be one of a life cycle costing model (LCCM). Although the term is of recent vintage, the concept is not. As indicated in Chapter I, it has long been recognized that all elements of a system, from R&D through production, operation and maintenance, and salvage or disposal must be taken into account. One practitioner suggests that it goes back to biblical times, citing Luke 14:28, "For which of you, intending to build a tower, sittith not down first, and counteth the cost, whether he hath sufficient to finish it?"[1] More recently, pioneering work was done by George Terborgh of the Machinery and Allied Products Institute.[2] From 1949 to 1967, with perhaps clear self-interest in promoting the sale of machine tool, Terborgh nevertheless made a significant contribution in pressing the point that most users of machine tools in the United States tended to underestimate the cost of maintenance and to retain them in use far too long. He therefore developed formulas for the costing of machine tools over their full life, including maintenance and repair, and discounting the cost stream. He also emphasized what in the military is called effectiveness by taking into account obsolescence and the greater productivity of alternative systems available later in time.

Nevertheless, with this bow to our ancestors and recognition that there is nothing new under the sun, it is worth a more detailed look at the history of why the Department of Defense came to place a much greater emphasis on life cycle costing in the late 1960s and the 1970s. Under the pressures noted above, over-runs on major military systems began in the 1960s to attract great attention and considerable concern in the Congress (a concern, needless to say, rapidly transmitted to the Department of Defense). Figure 1 gives some examples of these overruns. The figures represent an attempt to isolate the increases due to inflation. Homebuilders as well as weapons system buyers are familiar with the phenomenon of system growth, which means cost growth, as one adds extra rooms and frills to the house or additional capabilities to the weapon because they are feasible and nice to have.

There is also the ever-present possibility that contractor costs were underestimated for selling purposes. These factors have presumably been adjusted for in the figure. The systems used in the example in Figure 1 reflect to a large degree the fact that most of the major systems produced, with overruns, in the late sixties had been developed and costed in the fifties and perhaps early sixties, when prices were stable and changes in inflation rates were of little significance.

Because of major cost overruns in the 1960s (Lockheed C-5A and others) and the attendant unfavorable publicity, the Department of Defense focused first on curbing production costs. In July 1971, DoD issued Directive 5000.1, introducing the concept of "Design-to-Cost" (DTC). Thus, the attempt to reduce production costs started in the design, or development phase of RDT&E. The principal feature of DTC is that cost is to be given equal weight with performance as a design criterion, rather than being left as a dependent variable. An attempt is made to establish at the outset an affordable unit production cost (UPC) as a target against which contractors must seek the best performance possible within the cost ceiling. While it may not always be possible to adhere to the original cost target, cost be-

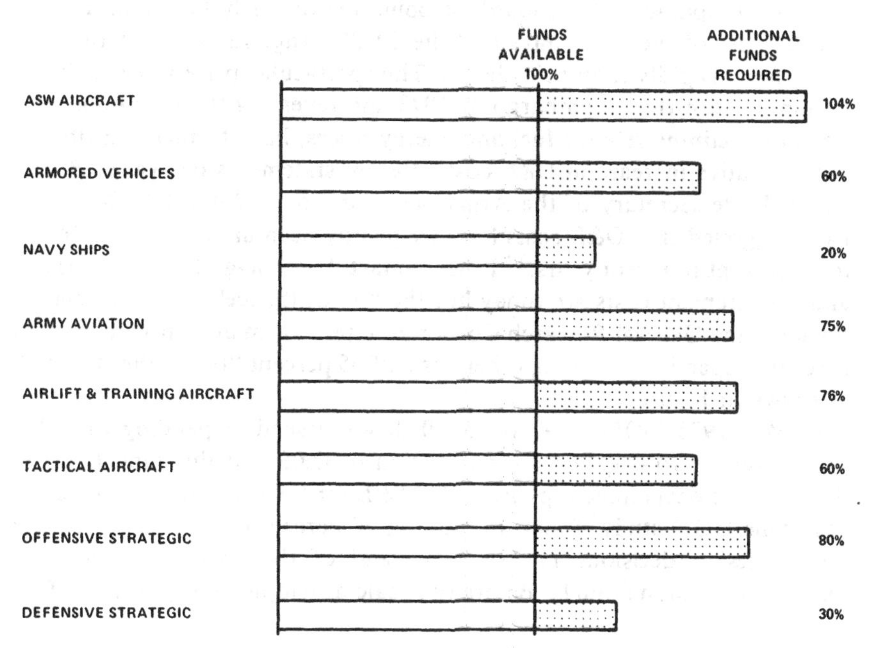

FIGURE 1. Inflation Impact

came a determinant of systems acquisition at least equal in importance to performance and schedule.

Cost control was to be exercised through modification or trade-offs. "Cost avoidance" became a by-word—what can a system do without, or what innovations can be made to avoid cost and lower the ultimate price of a system? An example is the designing of interchangeable wingflaps for the right and left wings of the F-16. In the F-16 case, some 450 trade-off studies were conducted and resulted in a projected, claimed cost avoidance of half-a-billion dollars over the life of the aircraft.

It rapidly became apparent, however, that DTC focused too strongly on production costs only. R&D and production costs have very high visibility, while the "costs of ownership," i.e., the costs of operations and maintenance (O&M) or operations and support (O&S) over the life of the system tend to be less conspicuous. But, since they go on for many years, they may be more important. Historically, O&S for weapons systems have constituted 50 percent or more of the total life cycle costs. This proportion has been rising over recent years, however, and has been estimated to average closer to 70 percent today. There are several reasons for this. Beginning in the late 1960's, military pay began to be increased quite rapidly toward the "civilian comparability" standard introduced in the early 1970's. Moreover, during most of the late 1960's and the 1970's, wage rates in general were rising more rapidly than price levels. The spectacular price levels of fuels, beginning with the oil embargo of 1973 and reversing the very long-term trend of declining relative fuel and energy prices, have further contributed to the relative increase in O&S costs. Recent statements by Walter B. LaBerge, Undersecretary of the Army, and Air Force Colonel K. M. Olver have suggested that O&S costs have frequently been underestimated by up to 50 percent in recent years.[3,4] It has, in fact, been suggested that RDT&E and Procurement costs are today but the "tip of the iceberg". A recent estimate of the General Research Corporation puts them at 10 percent and 25 percent, respectively, leaving other costs of 65 percent "below the surface", as shown in Figure 2.[1]

In May 1975 DOD Directive 5000.28 was issued, expanding the 1971 DTC directive to emphasize the inclusion of O&S and the importance of design in the development phase of RDT&E for O&S as well as production cost. One DOD study[3] found that about 70 percent of life cycle costs are set by design decisions made by the end of the concept studies (the DSARC-I decision point†). Before full-scale development begins (DSARC-

†DSARC: Defense systems Acquisition Review Council.

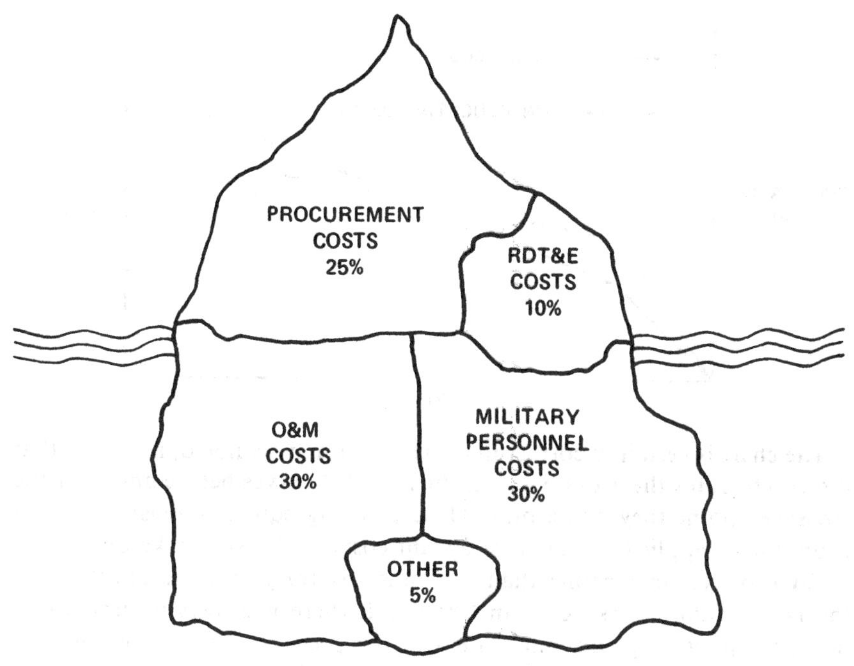

FIGURE 2. The Iceberg Effect

II), about 85 percent of the life cycle costs are locked up even though about 3 percent has been spent. By the end of full-scale development (DSARC-III), when the go-ahead is given to start production, 95 percent of the life cycle costs are frozen and cannot be changed except by expenditure of more money for redesign.

In short, there may be more opportunities to save money in the long run by reducing O&S than production costs. An example has been given by Grumman in the case of an aircraft for which the initial designs specified four different types of latches. After considering life cycle costs, including spares, storage and transportation, maintenance and replenishment, it was found that standardizing the latches did not reduce unit costs appreciably but affected an estimated savings of 30 percent. Going one step further, increasing the MTBF (mean time between failures) from 600 to 6000 hours increased unit price from $10 to $26 but led to an 80 percent reduction in life cycle costs in latches. Determining the optimum trade-off point between saving in production and in maintenance cost depends in part on the expected life cycle of the system in question, or the "time horizon" for which one is designing, as shown in this conceptual chart:[5]

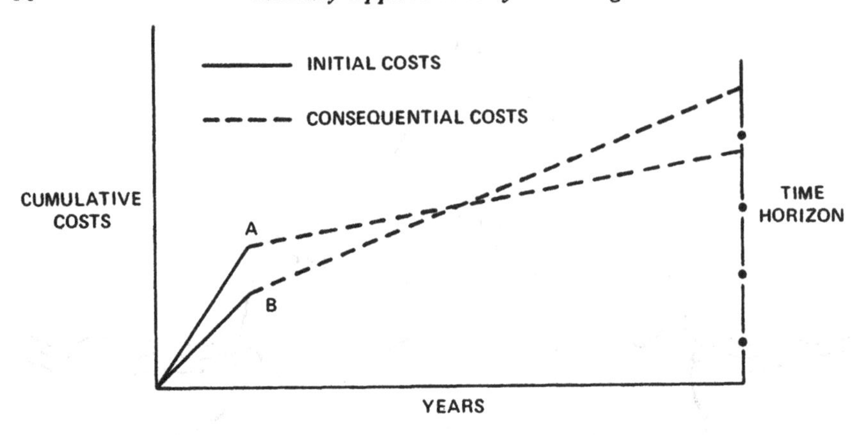

The chart is certainly correct in principal. It may be noted, however, that if one integrates the areas between the A and B curves before and after the crossover point they come quite close to being equal. A reasonable discount factor applied to the more distant costs might well make one decide in favor of system B rather than system A, contrary to the implications of the DoD example. As noted in Chapter I, there is always a human tendency to apply a high discount rate, which is precisely why the discipline of DoD directives and contracting procedures, forcing careful examination of the trade-offs, is necessary. One source has suggested that this human tendency is particularly manifested in the organ that controls the DoD budget, namely, the Congress; it is suggested[6] that Congress:

> behaves as if the discount rate were very high but does not explicitly state this position. When given the choice between spending one dollar today or spending much more in the future, it often defers the spending. This implies that members of Congress have a high internal discount rate; some have estimated this rate, as judged by their behavior, at 50 percent.

Nevertheless, the continuation of current trends for the increase in the relative costs of O&S will undoubtedly put pressures on Congress as well as DoD, and we may expect to see increasing emphasis on life cycle costing that includes O&S along with RDT&E and Procurement.†

---

†We have noted earlier, and will see in the ensuing example, that disposal and salvage costs are frequently ignored. This is certainly likely to be true in most DoD and Congressional decision stages. An interesting exception is to be found in nuclear weapons, in which the highest value element, the nuclear material, is subject to recovery and reprocessing; here, the salvage value, net of costs, clearly *cannot* be ignored.

An important caution is in order here. For many—perhaps most—systems it is not possible to make accurate forecasts of operational life span, about which decisions may be made far in the future. The real life span of the system (any system—think about your car or your house) is a function of many factors. These include rate and circumstances of use, rate of obsolescence (does something "better" come along?), rate of delay in obsolescence by the introduction of improvement "MODs," or modifications, and possibly even changed Defense budgets levels and other political factors. An extreme example that comes to mind is that of the B-52 and B-58 bombers. The two bombers were both conceived in the late 1940s and developed in the 1950s. The first B-52s were deployed in the late 1950s, the first B-58s in the early 1960s. For a number of reasons, but most importantly because improvements in high-altitude surface-to-air missile interceptors forced a revision in bomber doctrine from high-altitude to low-altitude penetration, the B-58 was phased out by 1970, whereas late model B-52s (considerably modified from the early models) are still in service and are expected to be kept in service through the 1980s. In short, expected life, or the time-horizon of the model, is perforce a somewhat arbitrary estimate (as are the estimates for R&D and Production costs for possible MODs). The best analysts can do is to try to identify whether there are any foreseeable reasons why one of two or more alternative systems should have a longer life than another.

THE TRI-TAC/U.S. MARINE CORPS LCCM PROJECT

To illustrate the problems and some methodological solutions of life cycle costing, we will review an ongoing, essentially state-of-the-art computerized life cycle cost model (LCCM). The TRI-TAC LCCM was initiated in 1978 to replace the original 1976 version developed by the Joint Tactical Communications Office (nicknamed TRI-TAC, for Tri-Service-Tactical) at Fort Monmouth, New Jersey. The TRI-TAC office conducts the Cost-Effectiveness Program Plan (CEPP) for joint tactical communications in conjunction with the Armed Services and Agencies of the U.S. Department of Defense. The 1978 TRI-TAC LCCM is currently being adapted, modified, and tested by the General Research Corporation (GRC) of McLean, Virginia, for use as the designated LCCM for a suite of eight tactical command and control systems envisioned for deployment from the mid-1980s through the end of the century. Named MTACCS (Marine Tactical Command and Control Systems), these eight systems are being conceived and developed by the Marine Corps, although some of them may become Joint Service ventures with the Air Force and Army, due to the integrated functions they would serve.

The Marine Corps issued an RFP (Request For Proposal) for an LCCM for its MTACCS systems in April 1978. At that time, the systems were in varying stages of R&D, with one awaiting approval to move into production. The Marine Corps issued a set of requirements for the MTACCS LCCM. The general purpose of the Life Cycle Cost Model effort would be to address all life cycle costs associated with development, procurement (production), operations, support, and disposal of MTACCS hardware and software. The model was expected to map the relationships between MTACCS cost elements and to be supported by a data base reflecting Marine Corps-unique force structure and manpower characteristics, especially manpower procurement, training, productivity, and turnover factors. It was expected to provide a measure of uncertainty regarding "best estimates" or expected output values and to provide the capability for both budget- and decision-relevant reports.

Further, the model was to be interactive, meaning that the model user would be able to communicate with the model on an immediate basis during its execution by the computer. The model would be modular and easily modifiable for each of the MTACCS systems. It was to be easy to update, and fully documented. Cost Estimating Relationships (CERs) embodied in the model should be easily understood. These requirements became the model design criteria used to select the LCCM for MTACCS.

The TRI-TAC model selected was designed specifically for comm-elec (communications electronics) systems, based on MIL-STD-881A, and has a three-level maintenance concept that conforms to most MTACCS maintenance plans. It is designed for Joint-Service use and contains manpower cost data for all the Services. It is extensively documented with a user manual. TRI-TAC has also been approved by CAIG (Department of Defense Cost Analysis Improvement Group) for DSARC presentations, and the TRI-TAC cost structure was developed in close coordination with CAIG.

While the TRI-TAC model was accepted as adequate for gross estimates during the concept development stage, it was judged inadequate for design trade-off analyses in the production phase. Accordingly, GRC recommended to the Marine Corps that three other models be used as adjuncts to TRI-TAC, to increase the reliability of the estimates.

The three adjunct models are PRICE, LOGOP, and GRC software. The PRICE model is used to determine R&D and production costs of electronics hardware. Output values from PRICE serve as direct input to TRI-TAC for contractor R&D and production costs that are hardware-related.

PRICE does not cost government expenditures. LOGOP can model equipment operations, failures, and logistics support. When used with TRI-TAC it can answer such questions as:

- Who should maintain the equipment—operators, technicians, both?
- How is equipment maintained—repair, replacement, or both?
- Where and in what quantity should maintenance resources be located?
- Does the logistics concept minimize maintenance personnel requirements?

The GRC software model is a parametric model for developing software estimates. Its outputs include costs to design, code, test, and debug software products.

Two points about the TRI-TAC hierarchy may be made here. First, the life cycle is not carried all the way through the disposal/salvage step. The common finding, noted earlier, was repeated here: disposal costs and salvage values are expected to net out to small enough amounts that, when discounted, they would represent an insignificant part of present value. Secondly, while TRI-TAC per se is deterministic, the Monte Carlo technique is used in LOGOP to randomize certain maintenance and operations events.

At the present time, the three second-tier models are interfaced with TRI-TAC manually. One would hope that in time the hierarchy of models would be made machine compatible, so that operator interaction would be feasible but manual interfacing would be unnecessary. (It is understood that they use FORTRAN as a common language.) As will be seen below, the number of cost categories and amount of data fully warrant computerization, and it would seem that this computerization should be complete in order to facilitate the ease of conducting many sensitivity tests and trade-off analyses.

Provision has been made for off-line work on O&S costs with the use of an HP-97 desk-top calculator. The program has been encoded on magnetic cards for this calculator, and the user puts his own data on additional cards. This has proved to be more economical than going on to the computer for some of the preliminary work with the O&S cost inputs.

Constructing the Model

The basic logic of the TRI-TAC model is shown in Figure 3. (It will be noted that the last two boxes go a step beyond the LCCM itself and imply the use of the costs in a cost-effectiveness analysis. The measurement and

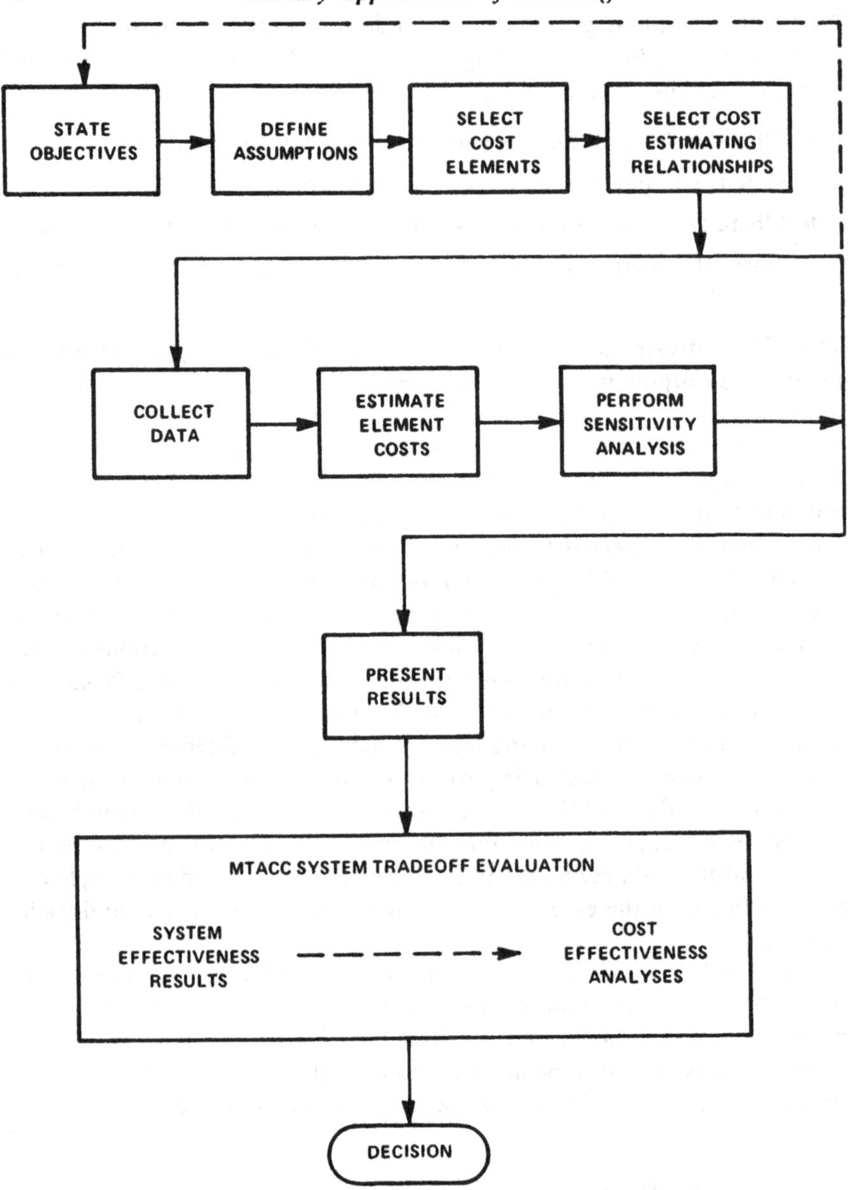

FIGURE 3. General LCC Methodology

analysis of effectiveness is beyond the scope of life cycle costing and of this case study. Nevertheless, it is important that cost analysts keep in mind this ultimate purpose of their costing.)

Having stated objectives and defined assumptions in close collaboration with the client, the primary organization step is then the third one, selecting the cost elements, or creating a cost-breakdown structure (CBS). The CBS is a structured array of cost elements defined first in terms of life cycle phases then in terms of end items. It is designed to encompass the entire set of costs incurred during the item's life span. The purpose of this perhaps tedious phase of the work is to ensure first that no items are omitted and second, and equally important, that there is no double counting by the inclusion of some costs in more than one cost element category. The CBS is what is often been called in accounting books and costing texts the "chart of accounts". In the present case, it starts at the first breakdown among the life cycle phases of R&D, production, and O&S. The next breakdown under these categories is between contractor- and government-incurred costs and the next binary split is between nonrecurring and recurring costs.

This top-down, or deductive, approach is supplemented by what might be called a bottom up, empirical, or synthetic approach of ordering all identified items in a logical fashion, which should in effect meet the analytical model half way. In this way, it will be found that some boxes do not have to be filled in as a perfectly logical structure might suggest. For example, in the O&S phase, contractors are largely out of the picture, although they may be involved on a minimal basis in training, revising software, or redesigning (value engineering). The completed CBS serves as a format for structuring the data bank and the model itself (Figures 4, 5, and 6). In the CBS for the 1978 TRI-TAC LCCM, it will be seen that R&D and production follow the binary splits mentioned above whereas the O&S phase is divided first into operations, logistic support, and personnel training and support. The columns in the tables provide for succesive aggregation of what is called Level 3 to Level 2 to Sub-element to Element. The final column, Category, is for the basic total of the major phases (R&D, production, and O&S). This columnar breakdown follows the work breakdown structure (WBS) originally established in MIL—STD 881A. An innovation here is the inclusion of personnel costs in the O&S phase. Most studies heretofore had excluded the costing of personnel by system or had set personnel apart as a separate category as in Figure 2 above. (For a discussion of the significance of MIL-STD 881A and the WBS, see 6.)

The CBS tables are highly summarized. In back of them lie thousands of data points and estimates. Collection of these data may amount to ninety percent of the LCCM effort.

TRI-TAC LIFE CYCLE COST ELEMENT STRUCTURE		COSTS IN (K) OF CONSTANT 19__ $				
	REGISTER NO.	LEVEL 3 WBS	LEVEL 2 WBS	SUB-ELEMENT	ELEMENT	CATEGORY
100 RESEARCH & DEVELOPMENT	R200					
110 CONCEPT FORMULATION & VALIDATION	R201					
111 CONTRACTOR	*R202					
112 GOVERNMENT	*R203					
120 FULL SCALE DEVELOPMENT	R204					
121 FULL SCALE DEVELOPMENT (NON-RECURRING)	R205					
122.1 CONTRACTOR (NON-RECURRING)	R206					
122.11 PRIME MISSION EQUIP (PME)	R207					
122.111 SUBSYSTEM (LIST COSTED SUBSYSTEMS)	*R208					
122.12 SYSTEM/PROJECT MANAGEMENT	R209					
122.121 SYSTEM ENGINEERING	*R210					
122.122 PROJECT MANAGEMENT	*R211					
122.13 SYSTEM TEST & EVALUATION	R212					
122.131 DEVELOPMENT TEST & EVALUATION	*R213					
122.132 OPERATIONAL TEST & EVALUATION	*R214					
122.133 MOCKUPS	*R215					
122.134 TEST & EVALUATION SUPPORT	*R216					
122.135 TEST FACILITIES	*R217					
122.14 TRAINING	R218					
122.141 EQUIPMENT	*R219					
122.142 SERVICES	*R220					
122.143 FACILITIES	*R221					
122.15 PECULIAR SUPPORT EQUIPMENT	*R222					
122.16 DATA	R223					
122.161 TECH. ORDERS & MANUALS	*R224					
122.162 ENGINEERING DATA	*R225					
122.163 MANAGEMENT DATA	*R226					
122.164 SUPPORT DATA	*R227					
122.165 SOFTWARE SUPPORT DATA	*R228					
122.17 OTHER (SPECIFY)	*R229					
123.1 GOVERNMENT (NON-RECURRING)	R230					
123.11 PROGRAM MANAGEMENT	*R231					
123.12 TEST SITE ACTIVATION	*R232					
123.13 GOVERNMENT TEST (DTE/IOTE)	*R233					
123.14 GOVT. FURN. EQUIP. (GFE) (SPECIFY)	*R234					
123.15 OTHER (SPECIFY)	*R235					
124 FULL SCALE DEVELOPMENT (RECURRING)	R236					
125.1 CONTRACTOR (RECURRING)	R237					
125.11 PRIME MISSION EQUIP. (PME)	R238					
125.111 SUBSYSTEM (LIST COSTED SUBSYSTEMS)	*R239					
125.12 SYSTEM/PROJECT MANAGEMENT	*R240					
125.13 OTHER (SPECIFY)	*R241					
126.1 GOVERNMENT (RECURRING)	*R242					
TOTAL RESEARCH & DEVELOPMENT COSTS						

N.B. ASTERISK IN REGISTER NUMBER COLUMN INDICATES TRI-TAC MODEL INPUT.

FIGURE 4. TRI-TAC CBS for Research and Development Costs

Data come basically from contractors and government sources. But they seldom "come easy". The cost analyst must identify the appropriate government agency or contractor, must ascertain whether the numbers exist at the level of disaggregation called for, whether the data are up-to-date, whether a need to know can be established, whether a letter of request, with various approvals, is required, and so on.

For new systems, cost estimates may simply not exist and must be made by the analyst. If the item is of any significance, it is usually necessary to determine parameters and equations that will give reliable estimates. These equations are often called cost estimating relationships, or CERs, of which

TRI-TAC LIFE CYCLE COST ELEMENT STRUCTURE	REGISTER NO.	COSTS IN (K) OF CONSTANT 19__ $				
		LEVEL 3 WBS	LEVEL 2 WBS	SUB-ELEMENT	ELEMENT	CATEGORY
200 PRODUCTION	R300					
210 PRODUCTION (NON-RECURRING)	R301					
211.1 CONTRACTOR (NON-RECURRING)	R302					
211.11 PRIME MISSION EQUIP. (PME)	R303					
211.111 SUBSYSTEM	R304					
(LIST COSTED SUBSYSTEMS)						
211.12 SYSTEM/PROJECT MANAGEMENT	R305					
211.121 SYSTEM ENGINEERING (NON-RECURRING)	*R306					
211.122 PROJECT MANAGEMENT	*R307					
211.13 TRAINING	R308					
211.131 EQUIPMENT	*R309					
211.132 SERVICES	*R310					
211.133 FACILITIES	*R311					
211.14 PECULIAR SUPPORT EQUIPMENT	*R312					
211.15 DATA	R313					
211.151 TECH. ORDERS & MANUALS	*R314					
211.152 ENGINEERING DATA	*R315					
211.153 MANAGEMENT DATA	*R316					
211.154 SUPPORT DATA	*R317					
211.155 SOFTWARE SUPPORT DATA	*R318					
211.16 INITIAL SPARES & REPAIR PARTS	*R319					
211.17 OTHER (SPECIFY)	*R320					
212.1 GOVERNMENT (NON-RECURRING)	R321					
212.11 INITIAL TRAINING	R322					
212.111 EQUIPMENT	*R323					
212.112 SERVICES	*R324					
212.113 FACILITIES	*R325					
212.12 SYSTEM TEST & EVALUATION	R326					
212.121 PRODUCTION ACCEPT. TEST & EVAL.	*R327					
212.122 OPERATIONAL TEST & EVAL. (OT&E)	*R328					
212.13 SYSTEM/PROJECT MANAGEMENT	*R329					
212.14 TEST SITE ACTIVATION	*R330					
212.15 COMMON SUPPORT EQUIP.	*R331					
212.16 SOFTWARE CENTER	*R332					
212.17 GOVT. FURN. EQUIP. (GFE) (SPECIFY)	*R333					
212.18 INVENTORY MANAGEMENT	*R334					
212.19 OTHER (SPECIFY)	*R335					
220 PRODUCTION (RECURRING)	R336					
221.1 CONTRACTOR (RECURRING)	R337					
221.11 PRIME MISSION EQUIP. (PME)	R338					
221.111 SUBSYSTEM	*R339					
(LIST COSTED SUBSYSTEMS)						
221.12 SYSTEM/PROJECT MANAGEMENT	R340					
221.121 SYSTEM ENGINEERING	*R341					
221.122 PROJECT MANAGEMENT	*R342					
221.13 OTHER (SPECIFY)	*R343					
222.1 GOVERNMENT (RECURRING)	R344					
222.11 QUALITY CONTROL & INSPECTION	*R345					
222.12 TRANSPORTATION	*R346					
222.13 OPERATIONAL/SITE ACTIVATION	R347					
222.131 SITE CONSTRUCTION	*R348					
222.132 SITE/SHIP/VEHICLE CONVERSION	*R349					
222.133 SYSTEMS ASSEMBLY, INSTALL. & CHECKOUT	*R350					
222.14 TECHNICAL ORDERS & MANUALS	*R351					
222.15 GOVT. FURN. EQUIP. (GFE) (SPECIFY)	*R352					
222.16 SUPPORT ENGINEERING	*R353					
222.17 OTHER (SPECIFY)	*R354					
TOTAL PRODUCTION COSTS						

N.B. ASTERISK IN REGISTER NUMBER COLUMN INDICATES TRI-TAC MODEL INPUT.

FIGURE 5. TRI-TAC CBS for Production Costs

more below. The magnitude of the problem is suggested by the fact that GRC has developed for TRI-TAC a text of some 500 pages, *The Data Collection Plan.* This volume serves as both a user's manual and data record book. For every cost element in the CBS, there is a minimum of two pages. On the left is the Cost Element Definition sheet, an example of which is

TRI-TAC LIFE CYCLE COST ELEMENT STRUCTURE	REGISTER NO.	COSTS IN (K) OF CONSTANT 19__ $ LEVEL 3 WBS	LEVEL 2 WBS	SUB-ELEMENT	ELEMENT	CATEGORY
300 OPERATING & SUPPORT	R030					
310 OPERATIONS	R031					
311 ENERGY CONSUMPTION	R033					
312 MATERIAL CONSUMPTION	R034					
313 OPERATOR PERSONNEL	R041					
314 OPERATIONAL FACILITIES	R050					
315 EQUIPMENT LEASEHOLDS	R051					
316 SOFTWARE SUPPORT	R087					
316.1 SOFTWARE PERSONNEL	R091					
316.2 SOFTWARE CENTER	R092					
317 OTHER OPERATIONS COSTS	R052					
320 LOGISTIC SUPPORT	R032					
321 MAINTENANCE	R084					
321.1 PERSONNEL	R035					
321.11 ORG. MAINTENANCE	R036					
321.12 INT. MAINTENANCE	R037					
321.13 DEPOT MAINTENANCE (OVERHAUL)	R038					
321.14 DEPOT MAINTENANCE (LRU/MOD. RPR.)	R069					
321.2 MAINTENANCE FACILITIES	R053					
321.3 SUPPORT EQUIP. MAINTENANCE	R039					
321.4 CONTRACTOR SERVICES	R054					
322 SUPPLY	R085					
322.1 PERSONNEL	R042					
322.11 ORG. SUPPLY	R093					
322.12 INT. SUPPLY	R094					
322.13 DEPOT SUPPLY						
322.2 SUSTAINING INVESTMENTS	R058					
322.21 REPLENISHMENT SPARES	R040					
322.211 ORG. LEVEL SPARES	R081					
322.212 INT./DEPOT LEVEL SPARES	R082					
322.213 REPAIR MATERIAL	R083					
322.22 MODIFICATIONS	R097					
322.23 REPLACEMENT COMMON SPT. EQ.	R098					
322.3 INVENTORY ADMINISTRATION	R043					
322.31 INVENTORY MANAGEMENT	R023					
322.32 INVENTORY DISTRIBUTION/HOLDING	R068					
322.33 TECHNICAL DATA SUPPORT	R059					
322.4 SUPPLY FACILITIES	R055					
322.5 TRANSPORTATION	R044					
323 OTHER LOGISTIC SUPPORT COSTS	R056					
330 PERSONNEL TRAINING & SUPPORT	R099					
331 REPLACEMENT TRAINING	R101					
332 HEALTH CARE	R102					
333 PERSONNEL ACTIVITIES (PCS)	R103					
334 PERSONNEL SUPPORT	R104					
335 BASE OPERATING SUPPORT	R105					
TOTAL OPERATING & SUPPORT COSTS						

FIGURE 6. TRI-TAC CBS for Operating and Support Costs

shown in Figure 7. The first line is an index number from the CBS, followed by a cost element title. The index number in the example indicates that the cost element is part of the O&S phase, more specifically, logistics support-supply. Five categories follow: a) definition; b) cost formula (model equation describing each variable in longhand); c) cost factors; d) model equation (citing the variables as they are labeled in the computer program; and e) user's note.

Facing each Cost Element Definition sheet is a Cost Element Data Collection sheet. On this page, all data and their sources are listed. A point estimate, or expected value, may be shown, or a high/low range, reflecting a degree of uncertainty. Given a range, an expected value may be indicated at other than the mid-point of the range. In many cases, sources may be reluc-

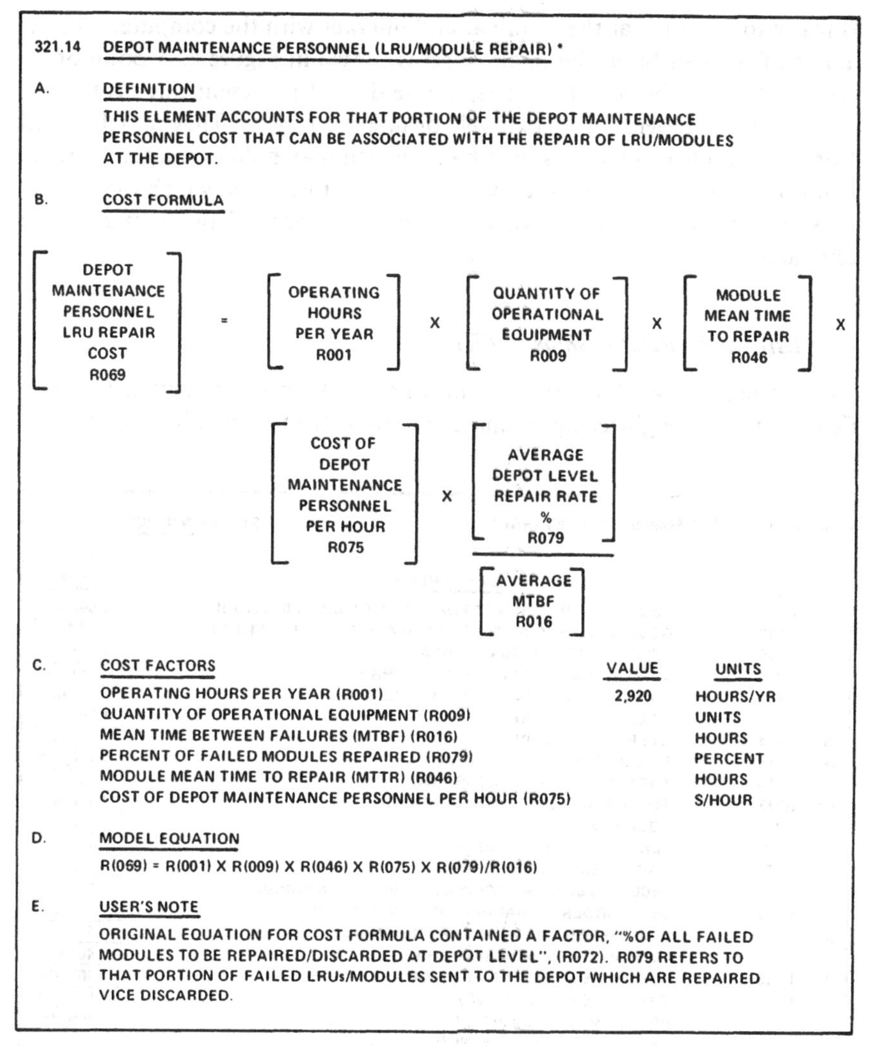

321.14 DEPOT MAINTENANCE PERSONNEL (LRU/MODULE REPAIR) *

A. DEFINITION

THIS ELEMENT ACCOUNTS FOR THAT PORTION OF THE DEPOT MAINTENANCE PERSONNEL COST THAT CAN BE ASSOCIATED WITH THE REPAIR OF LRU/MODULES AT THE DEPOT.

B. COST FORMULA

[DEPOT MAINTENANCE PERSONNEL LRU REPAIR COST R069] = [OPERATING HOURS PER YEAR R001] X [QUANTITY OF OPERATIONAL EQUIPMENT R009] X [MODULE MEAN TIME TO REPAIR R046] X [COST OF DEPOT MAINTENANCE PERSONNEL PER HOUR R075] X [AVERAGE DEPOT LEVEL REPAIR RATE % R079] / [AVERAGE MTBF R016]

C. COST FACTORS

COST FACTORS	VALUE	UNITS
OPERATING HOURS PER YEAR (R001)	2,920	HOURS/YR
QUANTITY OF OPERATIONAL EQUIPMENT (R009)		UNITS
MEAN TIME BETWEEN FAILURES (MTBF) (R016)		HOURS
PERCENT OF FAILED MODULES REPAIRED (R079)		PERCENT
MODULE MEAN TIME TO REPAIR (MTTR) (R046)		HOURS
COST OF DEPOT MAINTENANCE PERSONNEL PER HOUR (R075)		S/HOUR

D. MODEL EQUATION

R(069) = R(001) X R(009) X R(046) X R(075) X R(079)/R(016)

E. USER'S NOTE

ORIGINAL EQUATION FOR COST FORMULA CONTAINED A FACTOR, "%OF ALL FAILED MODULES TO BE REPAIRED/DISCARDED AT DEPOT LEVEL", (R072). R079 REFERS TO THAT PORTION OF FAILED LRUs/MODULES SENT TO THE DEPOT WHICH ARE REPAIRED VICE DISCARDED.

* LRU – LOWEST REPLACEMENT UNIT

FIGURE 7. A Sample Cost Element Definition Sheet

tant to give such data, knowing they will be cited. One useful ploy often used is therefore the entry of educated guesses by the analyst, upon which the source is asked to comment.

From the Data Collection sheet, an Input Summary sheet can be prepared. This Summary lists the input data in the order in which it will be called for by the model program. Once these sheets are filled out, the user is ready to sit down at the terminal and interact with the computer. An example of such an Input Summary sheet is shown in Figure 8. A point of interest is the number of .01 entries. These do not represent one-cent costs, but are place holders for unavailable or unknown costs. They may be filled from the established model's data bank, which may hold estimates for these items from other systems, or they may represent items on which the analyst must make future entries when he acquires more data or parametric estimates.

Cost Estimating Relationships (CERs)

As has been indicated, many items may not be defined with sufficient specificity to permit engineering estimates of costs. It may then be necessary to

LCC SYSTEM INPUT SUMMARY SHEET (R&D) SYSTEM MIFASS

	R #	DESCRIPTION	VALUE/ REF.
	R202	CONTRACTOR (CONCEPT FORMULATION & VALIDATION)	1,592,357.
	R203	GOVERNMENT (CONCEPT FORMULATION & VALIDATION)	.01
	R208	SUBSYSTEMS (NON-RECURRING)	12,403,000.
	R210	SYSTEM ENGINEERING (NON-RECURRING)	1,957,233.
	R211	PROJECT MANAGEMENT (NON-RECURRING)	1,151,079.
	R213	DT&E (NON-RECURRING)	1,793,561.
(a)	R214	OT&E (NON-RECURRING)	1,454,887.
(a)	R215	MOCKUPS (NON-RECURRING)	1,454,887.
	R216	T&E SUPPORT (NON-RECURRING)	1,798,561.
(a)	R217	TEST FACILITIES (NON-RECURRING)	1,454,887.
	R219	EQUIPMENT	207,006.
	R220	SERVICES (NON-RECURRING)	414,012.
	R221	FACILITIES (NON-RECURRING)	.01
	R222	PECULIAR SUPPORT EQUIPMENT (NON-RECURRING)	105,157.
(b)	R224	TECH. ORDERS & MANUALS (NON-RECURRING)	407,674.
(b)	R226	MANAGEMENT DATA (NON-RECURRING)	407,674.
(b)	R227	SUPPORT DATA (NON-RECURRING)	407,674.
(b)	R228	SOFTWARE SUPPORT DATA (NON-RECURRING)	407,674.
	R229	OTHER (NON-RECURRING)	1,877,000.
	R231	PROGRAM MANAGEMENT (NON-RECURRING)	600,000.
	R232	TEST SITE ACTIVATION (NON-RECURRING)	.01
	R233	GOVERNMENT (DTE/IOTE) (NON-RECURRING)	2,300,000.
	R234	GFE (NON-RECURRING)	112,000.
	R235	OTHER (NON-RECURRING)	.01
	R239	SUBSYSTEM (RECURRING)	8,438,848.
	R240	SYSTEM/PROJECT MANAGEMENT (RECURRING)	.01
	R241	OTHER (RECURRING)	.01
	R242	GOVERNMENT (RECURRING)	1,500,000.

(a) #214, #215, #217 FIGURED AS 1/3 EACH OF REMAINDER OF $7,961,783 (R212) MINUS 2 X $1,798,561 (R213 & R216)

(b) #224 – #228 FIGURED AS 1/5 EACH OF $2,038,369 (R223)

FIGURE 8. A Sample Model Input Sheet

find historical parameters, or independent variables, of which the cost is a derivable dependent variable within acceptable limits of variability. An example for an airframe is given in 7:

$$C = Ae^{B(\log V)-D}WRST,$$

where

C, the dependent variable, is airframe development and design cost;
e is the base of the natural logarithms;
A, B, and *D* are empirically-derived constraints;
V = maximum aircraft velocity in knots at maximum power and 55,000-foot altitude;
W = airframe weight in tons;
R = the hourly pay rate of engineering manpower;
S = is a factor which takes on either of two values, depending on whether the aircraft is fixed wing or variable sweep wing; and
T = the fraction of the airframe which is titanium.

Historically, CERs have been used chiefly for the costing of hardware items. The TRI-TAC model has extended the technique to the O&S phase. It has been common to estimate O&S costs as a percentage of production costs. Due to the increased relative importance of O&S costs discussed earlier, parametric analyses of many of these costs have been attempted.

The general procedure for the development of a CER involves the formulation of hypotheses as to cause and effect relationships between independent and dependent variables affecting costs. Some of these variables may be costs in themselves. If cause and effect are not provable, good statistical correlations may still justify their use.[8]

The parameters should be ones that can be specified and measured easily. For instance, it might be highly useful in determining transportation costs to know precise distances between bases and stations and repair centers. However, such information may be unknown or subject to change. It may, however, be possible to derive average distances between points for specified types of equipment and transportation and to use these averages in a CER.

With the increasing complexity of equipment, maintenance CERs offer particularly complex examples. Maintenance costs will correlate with both design characteristics and logistics arrangements. Several statistical averages have been utilized:

MBTF —Mean Time Between Failures
MTTR —Mean active Time To Restore the part, equipment, or system to satisfactory operation after it has failed
MLDT—Mean Logistics Delay Time, such as waiting for spares to arrive

MADT—Mean Administrative Delay Time
MDT (= MTTR + MLDT + MADT)

Figure 9 is a TRI-TAC CER for intermediate maintenance personnel that utilizes MBTF and MTTR. (The "*R*" numbers in the cost formula and model equation are merely register numbers assigned for tracking variables in the program.)

It must be noted that a parametric cost formula is not necessarily a CER. The example just given is referred to as a CER in part of the documentation of TRI-TAC.[1] If the six factors on the right hand side of the equation are known from experience, then the equation is simply a cost formula. If, on the other hand, some or all of these factors are unknown but are estimated for the equipment on hand on the basis of some parameter not shown (e.g., MTBF as a function of complexity, i.e., number of parts in similar equipment for which there is adequate experience to establish a predictive equation that meets statistical tests of significance), then the formula is a CER.

A better example is one cited by the same source for costs of manual preparation. Several factors are cited that must be estimated on the basis of some measurement of the complexity of the new system in relation to historical systems of similar characteristics (perhaps also taking account of common elements used in the new system): the difficulty of the material to be written, the total number of pages to be produced, and the number of original charts and diagrams required. (The number of pages, charts and

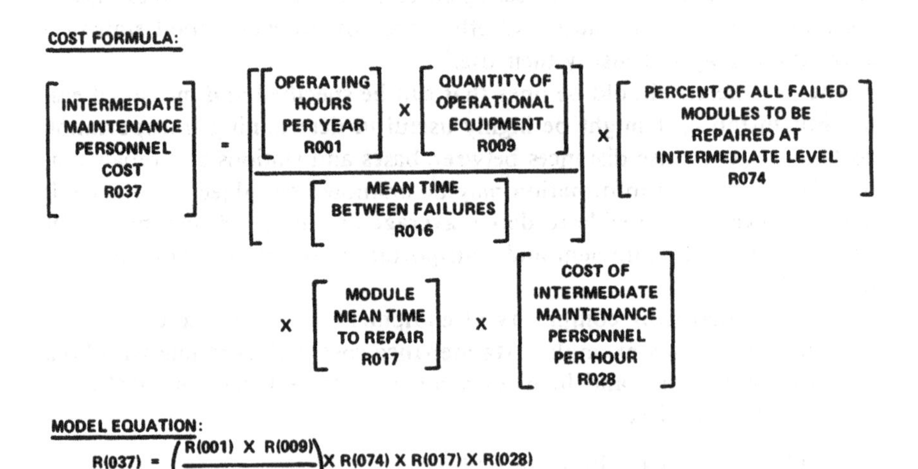

FIGURE 9. Intermediate Maintenance Personnel CER

diagrams may in fact be a function of the difficulty of the material.) Other factors taken into account are presumably more predictable or are policy options and therefore not dependent on CERs: physical page dimensions, paper quality, cover material, and quantity to be printed. That the problem is non-trivial is demonstrated by the statement that manual costs have been calculated to cost from $250 to $1,000 per page. While the numbers are not given, clearly, complex communication systems are likely to require large numbers of rather thick operating and maintenance manuals.

One further illustration showing greatly increasing complexity and the possible necessity for estimating on the basis of very limited experience arises from the increasing cost of commercial energy, plus the determination that alternative sources of energy should be provided, is illustrated in Figure 10, which gives the procedure for adding field electric power generators and both rechargeable and disposable batteries to the original formula for energy consumption costs as a function of average power rating times the number of operating hours per year times the cost of electrical power times the quantity of operational equipment.[1]

The Learning Curve

One CER of particular interest that has recently been added to TRI-TAC is that known as a "learning curve" that permits the calculation of unit production costs as a function of quantity of output. The learning curve theory (as applied to production costs) had its origin in World War II work associating the average effort, measured in man-hours, required for the production of aircraft, as the quantity increased. It was found that the average for a run of $2x$ aircraft would be eighty to ninety percent of the effort of x. This can be expressed by the formula

$$C = bx^n$$

where C is average cost, b is a constant (in dollars), x is the number of units produced, and n is a coefficient between 0 and -1. Thus, for an eighty-percent learning curve, sometimes more precisely called a "cost reduction curve," the formula is

$$C = bx^{-0.322}$$

This equation is linear on log-log paper. The model can use a different form of equation, where experience so dictates.

An interesting point may be noted about the learning curve. For any system, the total unit production cost will be the sum of the costs of many parts and subsystems. Unless all of these have identical learning curves, it

ORIGINAL COST FORMULA:

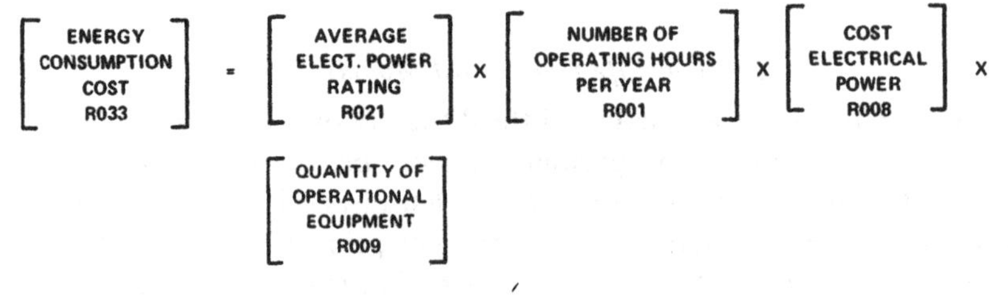

NEW COST FORMULAS:

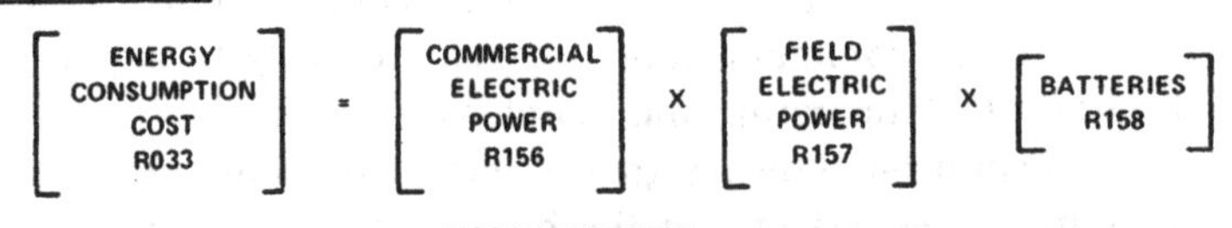

MODEL EQUATION:
R(033) = R(156) X R(157) X R(158)

COST FORMULA:

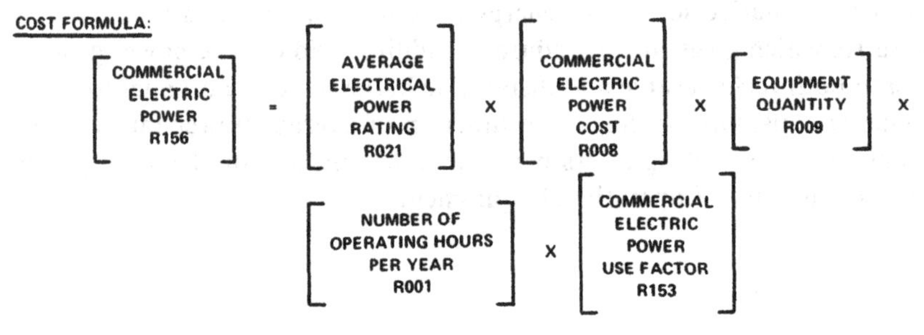

MODEL EQUATION:
R(156) = R(021) X R(008) X R(009) X R(001) X R(153)

COST FORMULA:

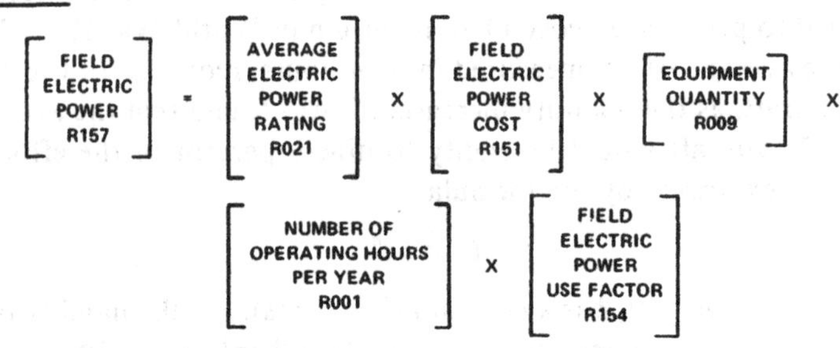

MODEL EQUATION:
R(157) = R(021) X R(151) X R(009) X R(001) X R(154)

COST FORMULA:

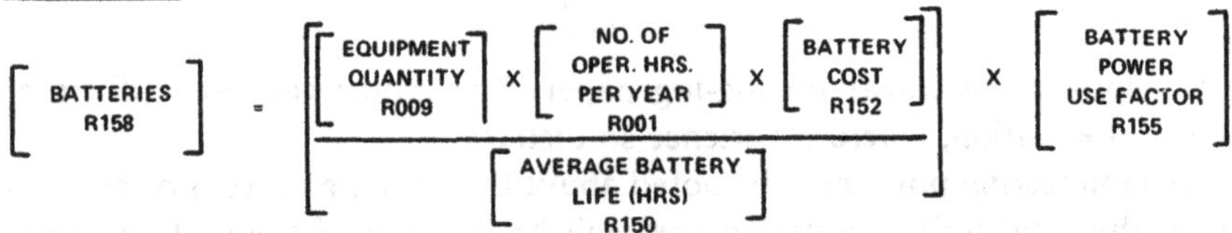

MODEL EQUATION:

$$R(158) = \frac{[R(009) \times R(001) \times R(152)]}{R(150)} \times R(155)$$

FIGURE 10. Energy Consumption CERs

can be shown mathematically that the sum of a series of costs derived by the above formula with different coefficients will be nonlinear, i.e., concave upwards; thus, the formula should be used with care for aggregate components or subsystems. For life cycle costs, this mathematical fact has a further interesting implication. One-time costs, such as R&D costs and initial plant and tooling, have a coefficient of -1 (a "fifty-percent" learning curve), that is, the formula becomes

$$C = ax^{-1}$$

where a is the total cost of the initial expenditure that is spread over x units. O&S costs, on the other hand, are often assumed to have very low learning curves (because, among other things, of labor turnover, particularly in military operating personnel). If we assume a "hundred percent" learning curve, (i.e., unit cost remains at 100% and learning at zero), then

$$C = cx^{0},$$

that is, the average cost $= c$ per unit and is constant.

This example is somewhat simplified. The phases into which the life cycle is divided are never really pure. For example: the recurring production of spares for maintenance will indeed involve some learning curve savings as quantities go up; R&D is in fact not likely to be a truly one-time cost, as, for any system maintained in the field for a significant number of years, some level of ongoing R&D is generally maintained to make improvements and occasional MODs, in the system.

To sum up,

$$C(\text{ave}) = ax^{-1} + bx^{n} + cx^{0},$$

where C(ave) = average (life cycle) unit cost, and

$$C(\text{total}) = a + bx^{n+1} + cx,$$

where C(total) = total life cycle cost. Both of these curves are significantly concave upward on log-log plots. Note that the curvature looks small, but this is because the log-log scales greatly compress the upper ends of both ordinates. Extrapolations can be risky. See Figure 11.

Rate of Output

One important variable in unit production cost not yet introduced into TRI-TAC is the rate of production. In general, if a larger plant is built to turn out a product at more units per unit of time, economies of scale are realized. In many fields, the cost of constructing and tooling a plant to produce $2x$ units per month might be 1.6 or 1.7 times the cost of a plant for

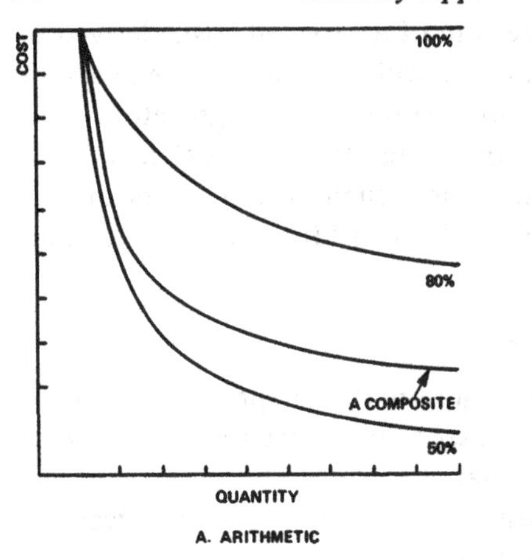

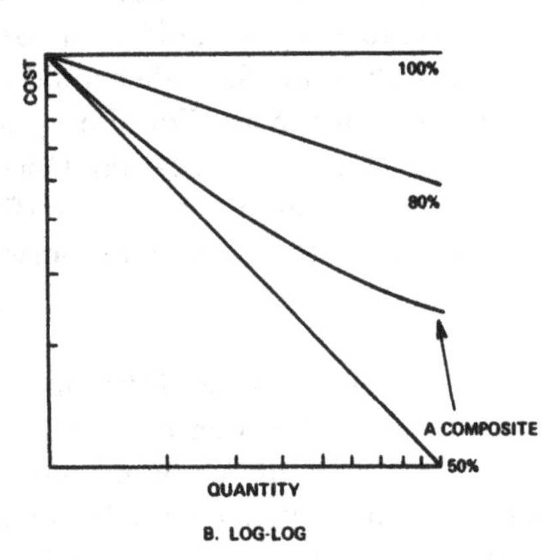

FIGURE 11. Sample Learning Curves

x per month. (The writer has no experience with the appropriate CER for electronic systems.)

Constant, Current, Discounted and Annual Costs

TRI-TAC currently has a capability for calculating costs in fixed, or constant dollars. It does this by "normalizing" all costs to the current year, inflating (more technically, "deflating") past years to the current year (although it is not clear why all or most such costs are not treated as sunk and irrelevant to the analysis) but not inflating future-year costs. It can generate a second set of costs in which past years are left in current dollars and future years are inflated in accordance with DoD guidelines. (We have already noted in Chapter I the deficiencies of then-year costing, but the present rules require that a DoD cost model be able to generate them.) The constant dollar series can also be discounted.

Also planned, but not yet introduced at time of writing, is a very important new feature, the capacity to generate annual costs, or costs by year incurred (presumably including O&S build-up between IOC and FOC). This step is important for several reasons. First, it will show if there are particular years of very high budgetary impact that might jeopardize the program politically. Second, it makes it easier to perceive the effect of change in the replacement time of a given system. Third, the presentation of annual costs enables a decisionmaker to separate sunk and future costs as of any future decision date he may wish to consider. Fourth, it provides the data for calculation of other cost formulations, such as "ten-year system cost".

The life cycle costs shown statically in Figure 7 are annualized in Figure 12—spread over time, as actually incurred. Here we see a long R&D phase preceding the big bulge for procurement. We also see that R&D may continue over most of the life of the system—in this hypothetical case, it led to one major MOD that required some additional procurement costs and extended the life of the system (and may have increased its effectiveness, or offset a decrease stemming from enemy defense improvements or countermeasures). O&S costs built up as the procured systems became operational, starting with initial operational capability (IOC) and leveling off at completion of deployment, or full operational capability (FOC) and, in this case, remaining constant until phase out. There was some net salvage value. Note again that, the higher the discount rate, the less the relative importance of the O&S and salvage value.

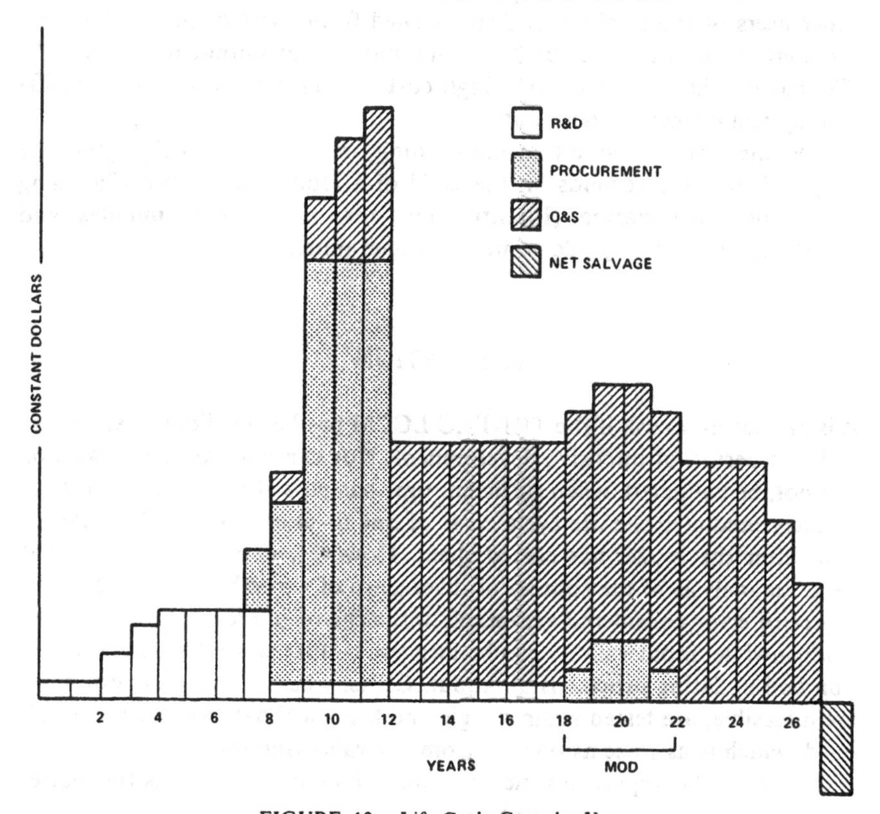

FIGURE 12. Life Cycle Costs by Year

Model Output

Since TRI-TAC is an interactive model, the user can control the outputs as well as the inputs. The previous section has described some of the output options now and soon to be available to the user. Unless otherwise instructed, the model starts its output with a recitation of the input equipments categories and values. This can be very useful in finding the source of any errors or the explanation of any seemingly unusual outputs. Output can then be specified for any level of detail, from total LCC, through life cycle phases, down to any level of disaggregation of categories and equipments desired. The model will automatically print out notes to the user on costs that were omitted for lack of sufficient inputs.

A particularly useful feature is believed by the system designers to be the capability to program the computer to convert totals of higher level cost categories in to columns on a bar graph. This visual presentation operates to save time and facilitate comprehension by clients (i.e., decisionmakers or other users of the findings as distinguished from users of the model, who are analysts who both input their data and format output of the model). The bar graphs may also make high cost drivers, i.e., particularly significant system elements, stand out.

It is interesting to note that the running time for the model is generally only 1/2 to 1 1/2 seconds on the CPU of a Honeywell Cyber 176, using Dartmouth time sharing. (Inputting may take ten to thirty minutes, and specifying outputs may add some additional time.)

VALIDATION

It is premature to ask if the TRI-TAC LCCM has been validated, since it is still in a period of intensive development. Nor can one say that it will, or will not, be validated, when sufficient time has passed to determine if it accurately predicted the life cycle costs of one or more systems that may be built after the model has costed them. However, if one could wait that long, an analysis of the sources of divergence of verifiable costs and of the degree of error considered acceptable would be of interest.

Meanwhile, input validity can be measured—that is what the data collection and updating process is all about. CERs, where engineering estimates are infeasible, are tested against logic and historical data as they are developed, which is as close as one can come to validating them in advance.

One particular aspect in which the input data are not valid is the matter of the inflation factors for calculating then-year dollars. We have argued that these are and will continue to be underestimated—but this is an institutional problem, not to be charged against the model.

Face validity would appear to be high, in that the logic of the model conforms to current theory as developed and accepted in a large body of literature (see References and Bibliography). It is also reported that TRI-TAC user acceptance is high—a common litmus test of face validity.

CONCLUSIONS

This case study has shown that the matter of costing military systems is complex and not precise, but both the theory and the practice are highly developed and the results can in general be made highly useful. A life cycle cost model can be, and the TRI-TAC model is:

- Highly flexible;
- Accurate enough for decisionmaking both among and within system programs, where performance or effectiveness can be adequately specified; system effectiveness must be estimated on a corresponding time-phased basis, or the LCC must be converted to a static total (say, 10-year system cost);
- Possibly accurate enough in an absolute sense to aid in high-level choices among non-comparable systems, or different parts of the military budget.

Costing is not cost-free. Several man-years of effort have gone into the TRI-TAC model, and this will continue, as might be expected in view of the fact that perhaps nine-tenths of the effort is in data collection. Certainly, the use of modern computers facilitates economy in the updating of data bases and the addition of new outputs as demand arises. The computer also greatly facilitates the calculations of tradeoffs and sensitivity tests.

The costs of costing are highly sensitive to the size and makeup of the costing teams. The requirements have been discussed by a number of sources. To cite but a few, Blanchard[9] advises that an analysis team should include some expertise in computer applications in modelling, experience in systems/product operations and logistics support, and reliability and maintainability, knowledge of internal producer and supplier activities, and of cost estimating and cost analysis. Seldon[6] lists no less than 13 disciplines required for life cycle costing: accounting, finance, estimating, engineering, manufacturing engineering, reliability, maintainability, quality control, logistics, management, statistical analysis, computer science, and contracting.

Analyzing the needs on a different level, Major Richard Grimm, USAF,[3] recommends that an LCC analyst have a technical background, high level analytical skills, imagination and creativity. Ford comments that "the most effective team is an amalgam of experience in different areas as suited for the particular life cycle costing project being undertaken. Engineers are not especially necessary for an LCCM effort whose goal is to cost only the O&S phase of the life cycle [this is really not life cycle costing—author]; they are vital for costing hardware development. Degreed academic backgrounds should not be overstressed at the expense of the desirable characteristics of analytical thinking, questioning, and problem-solving ability. Successful applications of life cycle costing rely on a fully integrated team complemented by active management support."[1]

The author, who has headed several costing teams, would find something to agree with in all of the above statements. Cost analysts want, and need, all the qualifications and help they can get. Still, the costing "teams" do not necessarily have to have experts in all the fields cited. The needs are not constant, and the use of consultants on specialties at different stages in the analysis and data collection may go far to meet the above objectives. The principal cost analyst, or team captain, may be an economist, engineer, operations researcher, or even something else. Certainly, he must be experienced. He must work closely and well with the customer for his model. In this respect, costing is not different from other kinds of military modeling. But it is an art that is well-developed and further along in solving major problems than most of those involved in the case studies in the remainder of this volume.

Questions

1. What can life-cycle costing contribute to the solution of weapons succession problems?
2. Can a cost model determine that a given system is "cost-effective?"
3. If one has a completed LCCM, why does he need annual costs as an output?
4. What is more rational, a high or a low discount rate?
5. How accurate does costing need to be?
6. Take a component of a type of system with which you are familiar and develop an hypothesis about a set of parameters on which data might be expected to be available and which might be expected to give a good estimate of the costs of the component in some future system.

References

1. Ford, Jerome C., *Life Cycle Costing: TRI-TAC/U.S. Marine Corps Life Cycle Cost Model,* General Research Corporation, McLean, Virginia 22102.
2. Terbourgh, George, *Dynamic Equipment Policy,* MAPI, Washington, D.C., 1949 and *Business Investment Management,* MAPI, Washington, D.C., 1967.
3. *Design to Life Cycle Cost,* AIIE Seminar, October 30–November 1, 1978, Washington, D.C., AIIE Seminars, P.O. Box 3587, Santa Monica, California 90403.
4. PRICE Users Meeting, Washington, D.C., April 23–25, 1979.
5. Department of Defense Guide LCC-1, "Life Cycle Costing Procurement Guide," U.S. Government Printing Office, Washington, D.C. 20402, July 1970.
6. Seldon, Robert M., *Life Cycle Costing: A Better Method of Government Procurement,* Westview Press, Boulder, Colorado, 1979, p. 128.
7. Department of Defense Guide, LCC-3, "Life Cycle Costing Guide for System Acquisition," U.S. Government Printing Office, Washington, D.C. 20402, January 1973, pp. 3–4.
8. Dodson, E. N., *Parametric Cost Analysis: General Procedures and Extensions,* General Research Corporation, P.O. Box 3587, Santa Barbara, California 93105, April 1976.
9. Blanchard, Benjamin, *Design and Manage to Life Cycle Costs,* M/A Press, Portland, Oregon, 1978.

Bibliography

AD 702-424, "An Introduction to Equipment Cost Estimating," Rand Corporation, Defense Documentation Center, December 1969.

AFR 173-10, "Cost Analysis—USAF Cost and Planning Factors," Department of the Air Force, February 6, 1975.

AFR 800-11, "Life Cycle Costing," Department of the Air Force, Headquarters USAF, Washington, D.C. 20330, August 1973.

AMCP 700-6 (Army), NAVMAT P5242 (Navy), AFLCP/AFSCP 800-19 (Air Force), "Joint Design-to-Cost Guide, A Conceptual Approach for Major Weapon System Acquisition," Department of the Army/Navy/Air Force, October 1973.

Clark, Rolf H., "Should Defense Managers Discount Future Costs?", *Defense Management Journal,* March 1978.

Collins, D. E., *Analysis of Available Life Cycle Cost Models and Their Applications,* Joint AFSC/AFLC Commander's Working Group of Life Cycle Cost, ASD/ACL, Wright-Patterson AFB, Ohio 45433, June 1976.

Department of Defense Directive 5000.28, "Design to Cost," Department of Defense, May 1975.

English, J. M. (ed.), *Cost Effectiveness—The Economic Evaluation of Engineering Systems,* John Wiley and Sons, Inc., New York, 1968.

Fabrycky, W. J. and Thuesen, G. J., *Economic Decision Analysis,* Prentice-Hall, Inc., Englewood Cliffs, New Jersey, 1974.

Fisher, G. H., *Cost Considerations in Systems Analysis,* American Elsevier Publishing Company, Inc., New York, 1971.

Gallagher, Paul F., *Project Estimating by Engineering Methods,* Hayden Books, New York, 1965.

Gillespie, C., *Standard and Direct Costing,* Prentice-Hall, Inc., Englewood Cliffs, New Jersey, 1962.

Goldman, Thomas A. (ed.), *Cost Effectiveness Analysis,* Frederick A. Praeger, Publishers, New York, 1967.

Haworth, D. P., "The Principles of Life Cycle Costing," Industrialization Forum, Volume 6, Number 3–4, Department of Architecture, Graduate School of Design, Harvard University, Cambridge, Massachusetts 12138, 1975.

Horngren, C. T., *Cost Accounting—A Managerial Emphasis*, 3rd Edition, Prentice-Hall, Inc., Englewood Cliffs, New Jersey, 1972.

Jelen, F. C. (ed.), *Cost and Optimization Engineering*, McGraw-Hill Book Company, New York, 1970.

Johnson, J., *Statistical Cost Analysis*, McGraw-Hill Book Company, Inc., New York, 1960.

Kendall, M. G. (ed.), *Cost-Benefit Analyses*, American Elsevier Publishing Company, Inc., New York, 1971.

Kernan, J. E., and Mencker, L. J., *Life Cycle Cost Procurement Guide*, Joint AFSC/AFLC Commander's Working Group on Life Cycle Cost, ASD/ACL, Wright-Patterson AFB, Ohio 45433, July 1976.

"Life Cycle Cost Reference Library Bibliography," compiled by A. Strofe, Joint AFSC/AFLC Commander's Working Group on Life Cycle Cost, ASD/ACL, Wright-Patterson AFB, Ohio, 45433, March 1976.

McClure, L., "Life Cycle Costing—A Selected Bibliography," RB 330-1, Martin-Marietta Aerospace Corporation, Orlando, Florida 32805, October 1976.

Menker, L. J., *Life Cycle Cost Analysis Guide*, Joint AFSC/AFLC Commander's Working Group on Life Cycle Cost, ASD/ACL, Wright-Patterson AFB, Ohio 45433, November 1975.

NADC-73240-50, "NADC Life Cycle Costing Methodology and Applications," Naval Air Development Center, Warminster, Pennsylvania 18974, November 1973.

Nanda, Ravinder and George Adler (eds.), *Learning Curves Theory and Application*, American Institute of Industrial Engineers, Norcross, Georgia, 1978.

NSIA AD HOC Committee Report, "Life Cycle Cost, Findings, and Recommendations," prepared for Assistant Secretary of Defense (I&L), April 1976.

Ostwald, P. F., *Cost Estimating for Engineering and Management*, Prentice-Hall, Inc., Englewood Cliffs, New Jersey, 1974.

Park, W. R., *Cost Engineering Analysis*, John Wiley and Sons, New York, 1973.

Petruschell, R. L., *Project Cost Estimating*, The Rand Corporation, Santa Monica, California, September 1967.

Popper, *Modern Cost Engineering Techniques*, McGraw-Hill Book Company, New York, 1970.

Quade, E. S., *Analysis for Military Decisions*, Rand McNally and Company, Chicago, 1964.

Quade, E. S. and Boucher, W. I. (eds.), *Systems Analysis and Policy Planning: Applications in Defense*, American Elsevier Publishing Company, Inc., New York, 1968.

Rudwick, Bernard H., *Systems Analysis for Effective Planning*, John Wiley and Sons, New York, 1969.

Seiler, C., *Introduction to Systems Cost Effectiveness*, John Wiley and Sons, Inc., New York, 1969.

Shisko, Robert, *Choosing the Discount Rate for Defense Decision Making*, The Rand Corporation, Santa Monica, California, 1976.

Smith, Arthur T. (ed.), *Economic Analysis and Military Resource Allocation*, Office, Comptroller of the Army, Washington, D.C., 1968.

Spurr, William A. and Bonini, Charles P., *Statistical Analysis for Business Decisions*, Richard C. Irwin, Inc., Homewood, Illinois, 1967.

Sutherland, William H., *A Primer of Cost Effectiveness*, Research Analysis Corporation, McLean, Virginia, 1967.

Trigg, Clifton T., *Guidelines for Cost Estimation by Analogy*, ECOM-4125, U.S. Army Electronics Command, Fort Monmouth, New Jersey, 1973.

TTO-ORT-032-76C-VC, "Cost Effectiveness Program Plan for Joint Tactical Communications," Fort Monmouth, New Jersey, April 1978.

Batchelder, C. A. (*et al.*), *An Introduction to Equipment Cost Estimating*, RM-6103-SA, The Rand Corporation, Santa Monica, California, December 1969.

"Bibliography on Design to Cost, Life Cycle Cost, and Cost Models," Defense Logistics Studies Information Exchange (DLSIE), U.S. Army Logistics Management Center, Fort Lee, Virginia 23801.

CHAPTER III

Strategic Mobility

THE PROBLEM

Strategic mobility is the capability to move forces to where they are required before or during conflict. Such movements may be *intertheater,* from CONUS, forward bases, or other theaters into the theater of conflict, or *intratheater*, involving the distribution or redistribution of military assets within the theater to the points at which they are or will be needed, generally from the rear toward the front.

The term "strategic" is used here in the classic sense of the overall planning for deployment of forces before and in a war, as against the "tactical" employment of forces in battles. The term is not used in the special nuclear-age sense of nuclear forces usable or used against the homelands of the antagonists. Thus, the focus is on General Purpose Forces for conventional (or tactical nuclear) warfare against the forces of the enemy.

This chapter will be concerned with intertheater mobility only and will focus on efforts to model the process of the strategic deployment of General Purpose Forces. The story starts in the early 1960s. The Kennedy/McNamara Administration was taking a "new look" at worldwide defense requirements. In reaction to the post-Korean Eisenhower Administration reliance on the "bigger bank for the buck" approach of "massive retaliation" to deter potential enemy use of General Purpose Forces, the new Administration sought a flexible response doctrine based on strengthened conventional capabilities. With Soviet testing of the new president in Berlin and Cuba, the memory of Korea still fresh, and Indochina simmering ever since World War II, "worst case" contingency planning did not seem unrealistic. The 2½-wars scenario was developed: a "major war" in Europe and another in the Far East (history repeating itself), plus a minor, or ½, war somewhere else. In the Nixon Administration, under the political and economic pressures generated by the Vietnam war, this was reduced to a 1½-wars scenario, i.e., a major and a minor war.[1]

It may also be added that the widely accepted theory of the sixties, that any major war would in fact be a short one involving major nuclear exchanges between the U.S. and the Soviet Union is gradually giving way to a belief in longer and more complex scenarios for the contingency of nuclear war between the U.S. and Soviet Union. It is increasingly believed that past analyses have underestimated the incentives on both sides for escalation control; nuclear exchanges will not necessarily be immediately "all-out" but may well be phased, or staged, over periods allowing for negotiation for war termination, or nuclear war termination, between attacks. Such a war is apt to be preceeded and accompanied by conventional warfare in the theater where the crisis started (Europe? Mid-East?) and possibly in other theaters, and on the high seas. The Soviets, for example, might, early in the conflict, wherever it started, deploy its forces forward in Eastern Europe, on the Chinese border, and to the south toward the Middle East, for protective purposes, both in terms of intimidating and deterring potential opportunistic invasions by peripheral enemies and for the purposes of dispersing its military targets and positioning them for possible preemption in one or more theaters. Scenarios for long wars are no longer considered inplausible (or at least no more implausible than short-war scenarios, given that most scenarios for wars, including actual past ones, are implausible).

How should U.S. forces be deployed for such contingencies? Forces could be forward-deployed overseas, in the danger spots of Korea, Germany, etc. They could be deployed on islands, ships, or other forward bases closer to the potential theaters of conflict than CONUS but not in-theater. Equipment could be prepositioned in theaters of potential conflict, with men and supplies to be deployed when needed. Forces could be kept in reserve in CONUS, with plans for deployment overseas in crisis or war.

What should be the mix? And what would be the criterion? Should forces be "sized" and then a least-cost mix be sought for the deployment plans for these forces? Or should the capabilities of available transportation—ships and aircraft—be taken as given constraints and the best feasible performance, or unit deployment profile, be determined? It was the old question, should effectiveness be fixed and cost minimized, or cost fixed and effectiveness maximized? It will be seen that both exercises were undertaken.

Our case history of intertheater mobility modeling will cover not one model but the evolution of a long series of models over a decade and a half. We will review work started for the Army in 1964 by RAC (Research Analysis Corporation), since taken over by GRC (General Research Corporation, now a part of Flow General, Inc.). Several tens of person-years are estimated to have been expended in this continuing project on the model building, data inputting (but not collection), and analysis. And while the analysts have been learning, the evolution of the models has also been spurred by the availability of new modeling tools and improved computers.

THE INITIAL LINEAR PROGRAMMING (LP) MODELS

The work started for the Army in 1964 was first formally reported in September 1966.[2] The report states that the model had already been through several stages, but the report was delayed because in 1965 the model started to be used extensively in support of analyses of the requirements for airlift/sealift forces by the office of the Assistant Secretary of Defense (Systems Analysis)—SA, now PA&E (Program Analysis and Evaluation).[3,4] The first models were linear programs to determine least-cost solutions for fixed effectiveness, specified as equal time-phased delivery profiles for given tonnages of specified material and supplies to given theaters. The version of the model reported in September 1966 used 400 equations to specify:

—Aircraft and Ship Types. Note that the emphasis at the time was on analyzing the potentials of the C-5A and FDL (Fast Deployment Logistics ship). The former was, of course, deployed, and the latter was killed in Congress—*not* on the basis of the findings of the models.

—Prepositioning. This included on land in the theaters in question and on Victory ships, designated "floating depot ships".

—POEs (Ports of Embarkation) and PODs (Ports of Debarkation). In point of fact, only one POE, CONUS, was designated, thus not requiring mathematical specification. This gross simplification was a serious limitation of the model, to be removed later. Alternate PODs were also not mathematically specified, but were rather data inputs for separate runs.

—A fixed amount of cargo, defined as outsize (O) or regular size (R), the outsize cargo being that which could not meet a size constraint specified for given aircraft. While an airborne division has no outsize components, a major portion of an armored division's equipment is outsize. The outsize cargo often limits the delivery time for an armored division.

—Deployment System. The collection of transport vehicles, equipment stocks, forward bases and associated facilities. Specification of transport vehicles included transport-vehicle productivity, reflecting area, time phase, route, class of tonnage (R or O), range and speed.

—Deployment-System policy:

Time frame—dates of availability of new components. (The overall time frame studied was 1970–1980.)

Transport Vehicle options

Prepositioning options

Distances, expressed as travel time from POE to POD.

—System component cost data. For existing systems, only O&M costs were specified, R&D and investment being treated as sunk. For new systems, R&D and investment were included, and unit costs varied with the numbers to be procured for a given least-cost solution. All costs were ten-year system costs undiscounted.

—Plan Effectiveness. Equal time-phase deployment profile.

A significant omission from the first models, fully recognized at the time, was the vulnerability of the components—vehicles en route, prepositioned materiel, and materiel delivered to the theater.

Model Outputs

The principal output of the model was the total system cost of the least-cost combination of deployment means within the constraints and assumptions specified by the user. In addition, the system specification of this combination was provided in a listing of the components that make up the optimal system, with the quantity of each component, the number of each type of transport vehicle to be maintained or acquired, and the location and type of each forward base, including the quantities of each class of equipment composing the total tonnage pre-stocked there.

A secondary output of the model was a set of deployment plans, one for each contingency area. These plans were specified in terms of tonnage, separated into the class previously identified with respect to the lift capacity of the transport vehicles. The plan specified, for each deployment, the tonnage moved or contributed by each system component together with the route over which the tonnage moved. The RAC report states that: "Implicit in the plan is the capability to assemble in the area the tonnage equivalent of the required forces in accordance with the force-closure schedule specified in the deployment strategy. A great deal of additional detailed planning would be required to convert these data into anything resembling a conventional movement plan. The generalized plans produced by the model as a part of the optimal solution could, however, provide the basis for such detailed planning. In this regard it should be noted that the tonnage plan merely indicates one way each deployment could be made. Usually the plan will not be unique. In areas in which one or more of the individual phase requirements is not limiting with respect to the overall system capacity, there exist alternative plans that could be accommodated by the system and that might prove superior from a tactical or other standpoint."[2]

Sensitivity Analyses

A feature of the linear programming technique is the marginal values in the basic solution printout. These represent the decrease in system cost that would result if the external-constraints value were to be increased by one unit. While these marginal values are strictly local and cannot be extrapolated, a relatively small number of runs will indicate how sensitive these values are to small changes in the constraints. In addition, the printouts can be programmed to show the range of values of component costs within which the solution would not be changed. Since these sensitivity data are limited to a single activity or item, however, and cannot be extrapolated with reliability, it is also useful to program the model parametrically, to give results for a range of values of such parameters as the pre-stockage limit, forward-based costs, time-phasing, and the capacity of the system to accommodate simultaneous contingencies—in other words, removing the fixed-effectiveness criterion.

The work reported in 1966 reflected the limitations of the then available computer at RAC, the IBM 7040. A published IBM linear program was used. The running time for the 400-equation program was 20 minutes. Running time for linear programs increases essentially with the cube of the number of equations, varying slightly with the density of the matrix of the variables and with the degree to which an initial basis is provided for the start of the linear program search. Essentially, however, to go from 400 to 800 equations would mean going from 20 minutes to about 2 hours and 40 minutes per run.

THE THIRD GENERATION

In 1967 and 1968, the linear program models were further developed to expand their capabilities, taking advantage of third-generation computers (the IBM-360 series).[5-10]

In 1965, the model utilized 300 constraints for five contingencies. In 1966, with simultaneous contingencies and concurrent nonwar requirements (CCNW), the model had grown to 800 constraints.[5] In 1967, a new model, MATRA (Mobility and Transportation Resources and Allocation) utilized 2,500 constraints, or equations, and 5,000 variables, requiring use of the IBM-360/65 computer.[7]

While the model was still linear, derived least-cost solutions, and covered the 1970-1980 time frame, it added greater specificity in a number of parameters. It divided the deployment time into ten periods (10-day linear increments for $D + 60$, plus 30-day increments through $D + 180$). It added commercial aircraft and ships. It employed several alternative cost functions, including discounted system costs. POEs were at three CONUS locations rather than one.

This increasing complexity was such that the model was broken into four submatrices: System activity (definitions, availabilities); Peacetime activity (five world areas); Strategic deployment activity (four theaters and a minor contingency, the latter being of the order of one tenth as demanding in terms of tonnages moved); and CCNW activity (five world areas with requirements *during* strategic deployments). These submatrices permitted the orderly preprocessing, or management, of the tremendous amount of data now involved. FORTRAN programs were developed to generate approximately 40,000 of the formidable total of 50,000 coefficients involved in MATRA. This use of matrix generators permitted flexibility and experimentation that would not have been feasible without the automation of a large share of the data management.

MATRA was impressively complex and flexible. Yet it had only pushed outward, not solved the serious limitations of LP models—and simulations were not yet available. The limitations were of several types:

—Mathematical. Integer solutions could not be assured, and "two-and-one-half ships" is not a useful output. Nonlinear constraints must be linearly represented. Fixed charge costs (which are nonlinear in per-unit terms) were not adequately included.

—Computational. These limitations were a reflection of pushing the state of the art in modeling and software, within the limitations of then existing hardware.

—Representational. Writing linear equations to describe the real world becomes increasingly difficult as one tries to include more and more of the real world problems. It is also terribly demanding in terms of communication between the modelers and the decision-makers they seek to serve. (This is an eternal problem, and it should be noted that communication appears to have been good and continuous in the present case. Perhaps, when the task is complex and evolutionary, contractor-performed modeling facilitates such communication over the long haul.)

—Analytic. The basic analytic difficulty is that of defining an "optimum" solution. No single objective function can be found that adequately meets the decision-maker's needs—there is no unique costing function, no uniquely preferable trade-off of delivery times versus risks and delivered quantities in varying scenarios. Attempts at weighting multiple objectives or defining objectives sequentially are inevitably subjective.

—Over-sensitivity. The costs could be affected greatly by insignificant changes in the requirements; it was impossible to state precisely that one set of requirements was militarily better than another. The mobility forces almost always were sized and costed by the amount of airlift required to meet the requirements in the first few days of a contingency. A minor addition of a few hundred tons to the requirements could add hundreds of millions in cost, with no measurable military benefit.

THE POSTURE SYSTEM

The arduous process of developing LP solutions to the strategic mobility problems of the DoD planners reached its peak in POSTURE.[10-18]

POSTURE was not in itself an LP but rather "a set of computer programs that will accept a relatively small set of data, expand this data into a large 'matrix', or set of linear constraint coefficients, and extract a concise and meaningful report from an LP solution." POSTURE is not an acronym but reflects the preoccupation with logistic "posture" defined as "the state(s) of military materiel and troops which are required to meet postulated contingencies. The state(s) of men and materiel include location, quantity, and readiness status. Thus, one element of a Tactical Air unit could have its posture described by the following: Air materiel, 5,000 short tons sited at site X, ready for movement in the time interval $D + 5$ to $D + 10$."[10]

The basis for the development of POSTURE was the conclusion in late 1967 that "flexibility was of greater importance than further expansion of detail; i.e., a means of generating and regenerating the matrix automatically, quickly, and easily."[10] Thus, POSTURE could generate MATRA, but it could also generate other models. These models could emphasize problems of basing and deployment states of forces, alternative transport systems, or different operational concepts for military units. Different contingency sets and theater combinations could be easily instead of arduously investigated.

The contingency sets, resources, and operating rules were set by the primary user, OSAD (SA), and the RAC analysts conceived and generated

LP models, using POSTURE. Solutions of these models, and successive variations reflecting changes in the specified resources and rules, were then an interactive process between user and analysts to aid the user in reaching decisions.

The basic achievement of POSTURE was a high degree of automation of the generation of large LP matrices. The matrices were generated by FORTRAN IV on an IBM 7044 and then inverted and solutions printed out by an IBM 360/65.

The procedure started with preprocessor routines for both aircraft and ship input data. Only the Air Preprocessor will be described here. Inputs to the Preprocessor were in the form of a series of tables, as follows:

—Maximum number of routes; time basis for aircraft productivities, in day; maximum number of origin sites; and total number of periods in which aircraft may reallocate.

—Route numbers for each allowable combination of five theaters and nine origins.

—Type of load that can be carried by each aircraft type.

—For permitted loads (above), productivity adjustment ratios for mixed troop/cargo loads vs. all-cargo loads.

—Reallocation periods, in days (e.g., 5-, 10-, and 30-day periods).

—The last period in which each route will be used.

—The recovery time, in days, for each aircraft type (this being at most two days; aircraft, unlike ships, can be treated as one worldwide pool of available lift equipment).

—Adjustment ratio if aircraft utilization rate is to be changed, by relocating the time period for each aircraft type.

—Number of passengers per kiloton of cargo, in mixed loads, by aircraft type.

—Aircraft productivities, in terms of number of aircraft required per thousand units delivered, by aircraft type, by route, by load, by unit.

The output of the Air Preprocessor to the Matrix Generator was a table of aircraft productivities by aircraft type, route and loading (all troops, all cargo, mixed).

The Ship Preprocessor was similar, with the additional provision of troop and cargo assembly times, ship speeds, and loading and unloading times.

Fig. 13 shows the role of the Preprocessors and the flow of data through the Matrix Generator in the POSTURE system.

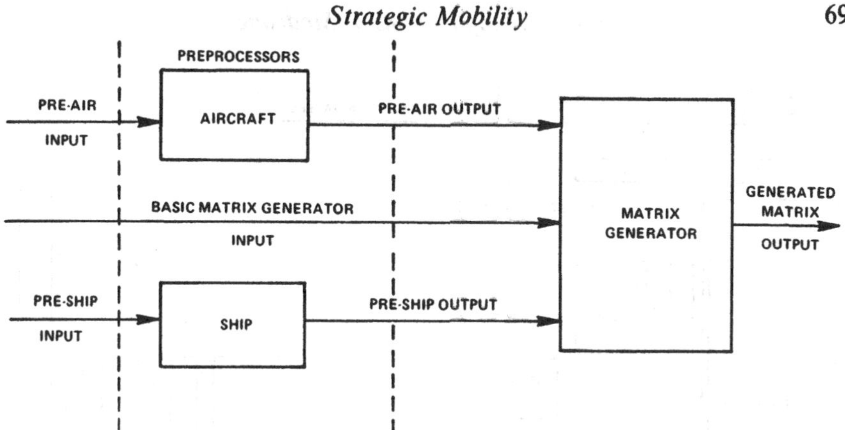

FIGURE 13. POSTURE Data Flow Through Matrix Generation

The Figure shows that, in addition to the data provided by the Preprocessors, there were direct inputs to the Matrix Generator. These were extensive. We will not try to detail them here but note that they covered: specification of contingency locations; availability of aircraft and ships (with locations of ships); resource (troops, equipment and supplies) readiness (time of availability); resource requirements in each theater; peacetime and new assets, undiscounted and discounted, basing cost, and readiness cost). Clearly, a tremendous volume of data was involved. But at any one time the user of the system would probably change only one or a few variables and/or constraints (constants, or "right-hand side"—RHS—entries in the equations that constitute the rows of a linear program matrix). The Preprocessors and Matrix Generator took care of the laborious process of translating these changes into new equations and an LP matrix. This matrix was called a POSTURE-generated model, and a schematic example is given in Figure 14 (based on [10]). The matrix was solved on an IBM 360/65 (or a comparable computer, the CDC 6400), generally in less than one hour running time; the output was formatted by a Report Writer program according to the user's wishes. The output was usually in the form of tables showing one or more cost functions (e.g., undiscounted and discounted) and numbers of aircraft and ships required, by type. Alternative models could be generated to show performance (delivery-time profiles) for fixed cost.

POSTURE continued to be improved and utilized for several years,[12-18] (and expecially 17). Three alternative formulations for force capability analyses were developed. MaxCap maximized total tonnage delivered by a given system in a given period. Profile maximized tonnage delivered subject to at most one time-phased profile per theater. RapDep maximized tonnage

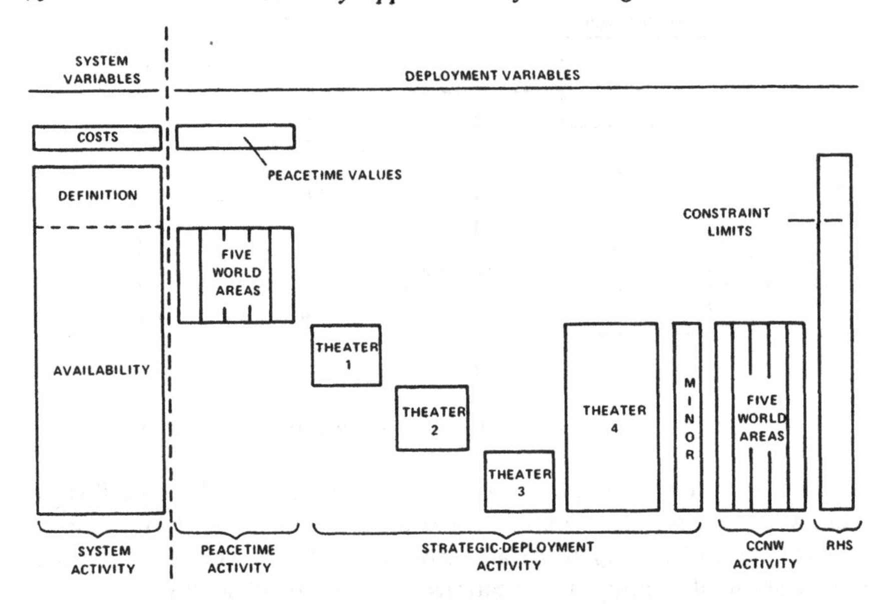

FIGURE 14. A POSTURE-Generated Model: Schematic Representation of Macrostructure

delivered in one or more specified time periods. A further development undertaken for the Logistics Directorate, J-4, of the Organization of the Joint Chiefs of Staff, was called PREPKG-POSTURE. This system achieved the purposes of the above three formulations but greatly reduced computation time by means of three modifications of the POSTURE system: (1) unit requirements were defined at each POD from each POE; (2) materiel availability at each POE was defined for each POD; and (3) the PREPKG matrix was substituded for the specific Right Hand Side (RHS) materiel requirements. The first two steps were designed to prevent POSTURE from using materiel at a nearby POE to meet requirements for units based at a more remote POE and from meeting unit requirements at one POD by using units destined for other PODs that happened to be available earlier in time than the units desired. The third change accounted for the reduction in computing time by substituting an objective function for the parametric analyses on the matrix Right-Hand Side.

A further step was the incorporation of a new IBM Mixed Integer Programming (MIP) algorithm. This algorithm permits the specification that certain variables, e.g., ships and military units, can take on only integer values. This prevented the 2.5-ships type of solution, permitted the imposition of minimum numbers of ships in convoys, and the maintenance of unit integrity of major combat divisions.

The introduction of convoys was a major step forward. For this purpose, the models degraded the performance characteristics of the convoy ships by increasing their sailing distances and decreasing their speeds while in convoy, since the convoy speed must be less than the maximum speed of the slowest vessel in order to permit stationkeeping. The criterion for decision as to whether to form a convoy was the effect on the overall deployment when it was completed. However, with Mixted Integer Programming, the convoy analyses took several hours of IBM 360/365 running time.

SIMULATION OBSOLETES LINEAR PROGRAMMING

In 1970, development began on a simulation model, QTYP (Quicker Than a Yellow Pad—presumably still slower than the back of an envelope). This model was a deterministic simulation which used a simple heuristic rule to schedule the movement of forces. The model was originated as an alternative to frequent use of POSTURE for small problems with short response times. The computer run time of a typical simulation was at most several minutes.

The objective of the simulation was to maximize the utilization of available transportation by delivering each of a set of forces and resupply as soon as possible. The forces were processed in priority order, where the priority was determined by a specified Required Delivery Date(RDD). The heuristic rule used in the model was: assign as much of a unit as possible to the transportation resource which will deliver a portion of the unit at the earliest date. Among the greatest advantages of this model were its low cost, fast running time, and the capability to represent the scheduling process in more detail. This model simulated activities on a daily basis, and more realistically.

An amusing side effect of QTYP illustrates the old adage, "Never be as clever as you can." Apparently someone in the bureaucracy thought the term, QTYP, a bit too cute—perhaps because it reminded some naval person of the vulgar appellation of "swab" for sailor. In any event, someone put in parentheses after QTYP "(not an acronym)," and the acronym died by fiat. New names, SDSS (Strategic Deployment Simulation System) and RAPIDSIM (self explanatory), were used for subsequent mods in the series. series.

The model grew even more sophisticated and was used with increasing frequency until it displaced the LP models it was created to complement. For example, it was used in 1973 for the Navy's Sealift Procurement and

National Security (SPANS) study and other DoD studies, and in 1974 in the airlift enhancement study, which included the C-141 stretch version and the 747 modified with a large side door for military use. Three categories of cargo were now specified: (1) bulk (palletizable for a Boeing 707 or larger aircraft); (2) oversize (C-141 or larger); and (3) outsize (C-5 only).

The basic difficulty with the model was that it was limited in the activities it could realistically model; the heuristic algorithm could not support many of the desirable refinements. The model did not simulate in time sequence; it did not increment a clock day-by-day while scheduling the activities which occur on that day. Thus, the model could not be used to assemble a convoy of minimum size unless it was run iteratively. Nor could it be used to select a single transportation mode for the movement of a cargo unless a single mode were specified or the model were run in an iterative mode. Running the model iteratively required the user to make decisions and adjust the model inputs after each run; this could be very time consuming.

Lastly, the model did not maximize the utilization of resources, since transportation resources could be reserved for long periods to move a unit not yet available to deploy but which had a high priority. The utilization of transportation resources is thus reduced, since the ship or aircraft could have delivered other cargo in the interim.

THE FOURTH GENERATION—ISDM

Beginning in 1975, a new order of realism was introduced into the OASD (PA&E) series of strategic mobility models. The limitations of both LP and third-generation computers for optimization had become apparent to all. But the fourth generation of computers was becoming available, and the modelers were ready to take advantage of it with a simulation that could trade the LP quality of optimal solutions for the simulation quality of detailed information about how a deployment could be carried out. The analysts had concluded that the decision rules in a logistics simulation could yield solutions acceptably close to the LP solution.

The Interactive Strategic Deployment Model (ISDM) was designed to make maximum use of the capabilities of a fourth-generation computer while maintaining a running time of fifteen to twenty minutes. An actual running time of two to three seconds of CPU (Central Processing Unit) time was achieved per day of simulation, or less than ten minutes for a 180-day simulation.

The fourth generation computer utilized was the Honeywell 6180 MULTICS at the USAF Data Services Center at the Pentagon. The MULTICS is a "virtual memory" computer that treats the entire capacity of the computer as a part of the memory core. The computer can detect whether a given "page" of information is in the memory, and if it is not the computer can put it there.

The term "interactive" in the name of the model does not imply man-machine interaction during the simulation, as this was found to be too time-consuming and costly. Rather, it is intended to emphasize the high degree of interaction feasible in inputting data in scenarios and selecting output, which can occur at any point in the simulation of its conclusions, in display, hard copy, or microfiche, including graphical as well as numerical representations. Using an information management program, the user can generate reports directly from his data bases or from the outputs of any simulation run. In addition, he may store the results of a simulation and retrieve them at any time for further analysis and additional reports.

To describe the ISDM, it is not necessary to start with all the inputs. It is sufficient to note that the above-described POSTURE system was a point of departure and to note some incremental capabilities. The most significant point is the considerable increase in flexibility, both in steps in the logistic process and in heuristic decision rules that can be incorporated. For example, ISDM can, if need be, simulate the use of port facilities, the facilities needed to load units at their origins and transport them to POEs, and the reassembly of units in the theater and their transportation to the battle area. At time of writing, these steps have not been added because they have not been judged worth the considerable increases in data requirements and running time, but the model incorporates user-supplied estimates of the times required for each of these activities.

Even more important, the model is designed to utilize the output of a sub-model of attrition at sea, that is, to take account of the exogenous effects of enemy action. Again, at time of writing, this feature has not been added, but the model accepts user estimates of this attrition. The reason for not incorporating the attrition calculations is not, in this case, the increase in running time, but rather the fact that Navy approval of the attrition factors has not been obtained. (Airlift attrition has therefore also been omitted.) Methodologically, however, and perhaps in the future practically, this capability is a great step forward. The simulation avoids a major problem of incorporating attrition in a linear program; it treats attrition as an uncertainty, in terms of shipping decisions, whereas an optimizing linear program in effect knows the attrition outcomes even though these are future events at the time of shipping decisions.

An overview of the simulation process may be of interest. ISDM is a scheduling model that uses a set of heuristic decision rules to make assignments of transportation resources to move cargos. The model schedules in priority order as many movements as it can on each day, trying to satisfy all the constraints; then it either schedules the next day or, if the current schedule violates a constraint, it backtracks and schedules a less desirable alternative. Backtracking is limited to one day. There is no attempt to improve the solution by backtracking over serveral days; backtracking more than one day may cause the higher-priority cargo moved on the earlier day to be delayed, thus violating the priority concept. Moreover, there is the above-mentioned element of uncertainty about future capability to deliver cargo due to the possibility of interdiction, and the cost of delay in warfighting capability could be large. The prudent choice is to make good use of the resources in hand.

The model in essence simulates the process a military planner might go through to generate a deployment schedule. Certain assumptions are made:

1) The planner has reliable data on the availability and composition of his forces.
2) The planner has reliable data on the numbers, location, and status of ships and aircraft.
3) The planner cannot know which ships and aircraft will be lost due to attrition.
4) The planner cannot know when or whether the enemy will actually attack. He must plan on the basis of his assessment of the threat of attack.
5) The planner has reliable intelligence about the deployment capabilities of aircraft and submarines which might interdict his lines of communication.
6) The planner can accurately predict the level of attrition of ships far enough in advance to assess the risks he is willing to accept. Therefore, he can modify his policy for scheduling independent ships and convoys to avoid unacceptable losses.

Note that the last two assumptions are rather heroic. They probably should be subject to as much study and parametric analysis as all the rest of the problem, but as of now they are simply exogenous to the model. The user can put in whatever specific assumptions he chooses.

There are also certain criteria for a good schedule:

1) It should make maximum use of the transportation resources. Ships and aircraft should not stand idle if cargo is available to move.

2) Each unit should be transported as quickly as possible. If there is a choice, resources should be assigned to get each unit to its destination at the earliest feasible data.
3) Forces should be moved in the order in which they are needed. For example, resupply should not arrive at a POD before the combat forces for which they are intended.

These objectives cannot be fully satisfied simultaneously by a single schedule. First, some forces are closer to POEs than others. Second, some forces may not be ready to move when needed. Third, there may be mismatches in the availability of forces and transportation resources. For example, container ships might be available but not containerizable cargo.

Given these assumptions and criteria, we may examine the modeling of sealift and airlift. ISDM tracks the movement of each ship on each voyage, and its cargo can be identified. The time at which a ship may start loading at each port depends on the initial position of the ship at mobilization or the time of return to each port after a previous voyage. The actual assignment of the ship is determined by the earliest delivery of a package, not the most efficient use of the ship.

Individual aircraft are not tracked. As noted in the earlier discussion of the linear programming models, the short flight time of any aircraft to any location permits great simplification of the data requirements and the shortening of run times by treating aircraft in terms of the productivity of each type of aircraft on each day. This aggregation, however, precludes the identification of queuing problems at overloaded air bases.

The model schedules as many movements (assignments of aircraft or ships to haul cargos) as possible on one day, repeating the process for each day of the simulation. There are many possible schedules which can be generated on a single day. Some schedules are preferred over others. For example, the model always prefers a schedule in which all ships sail independently over one where some of the ships are in convoys. Ships in convoys will arrive later. But the user may instruct the model to use convoys because of the assessment of risk to independent ships.

Scheduling decisions are made on four general levels, with a number of decisions being made on one level before a decision can be made at the next higher level. These levels are:

I. The schedule is tested for violation of any constraints imposed by the user. If a schedule is acceptable, the model will move to the next day. If a constraint is violated, the model must seek an alternative, presumably less preferred, schedule.

II. ISDM selects the highest priority package of forces that is available to move, has not been completely scheduled, and for which suitable transportation resources are available.

III. The model compares the potential delivery dates of the package via airlift and sealift and selects the mode which gives the earliest delivery (to the theater or to the FEBA). Ties are broken in favor of sealift. This is an arbitrary decision which in the long run does not affect the results.

IV. Level IV is a mechanism for generating the delivery dates used in Level III. Airlift and sealift are separate.

A. Assuming the package can deploy by airlift, aircraft sorties and utilization hours are allocated to achieve the earliest delivery of the package. Depending upon the complexity of this allocation and the preferred use of each type of aircraft to haul bulk, oversize or outsize cargo, the solution may utilize a linear program as a subroutine. (Up to 1,000 LPs may consume thirty seconds of a ten-minute run!)

B. Assuming the package can deploy by sealift, the model will seek the port through which the earliest completion of the delivery can be scheduled, taking the origin of the cargo and the availability of ships at each POE.

A typical simulation may involve from 10^9 to 10^{10} possible individual movements from origin to destination. The constraints and allocation rules, of course, greatly curtail the actual search. The numbers actually considered are also reduced by the "preprocessing" aggregation of many of the data base entries of similar items into single entries. The assumption behind the aggregation is that there is little significant information lost if units that have similar characteristics, similar origins and destinations, and the same availability and required delivery dates are treated as a single entity. The primary sorting parameter for aggregation is the required delivery date (RDD), this being the parameter that determines the sequencing of forces.

There are usually less than 10^4 individual movements in the acceptable schedule arrived at by the model.

The Time Clock for Scheduling

A movement is actually a set of events that must occur in sequence, e.g., start of loading of a ship, completion of loading, departure of the ship from the port, and arrival at the POD. An event has significance only in context; in isolation, its value cannot be determined, nor can a decision be made on its scheduling. It is therefore necessary to select a point of reference in the sequence of events. This point of reference is defined arbitrarily at some point between the POE and the POD. The time of arrival at the point of reference is the clock for the simulation. If the scenario includes attrition, the first exposure to attack by the enemy is the most useful point of reference, since attrition losses on a given day must be taken into account in the conditions at the start of the next day.

The determination of a point of reference is more difficult for aircraft, since they may complete one or more round trips in a day. Moreover, packages delivered by aircraft may depart many days later than ship-delivered packages and still arrive earlier. The ISDM solution is to define a point of reference in time, or a clock, for airlift which permits estimation of tradeoffs between air and sea delivery modes. This is generally a time such that a package arriving by air could reach its destination on the same day as it would if it arrived by sea.

If everything could be delivered by air, there would be no problem. Aircraft would always be preferred, and everything would be delivered sooner. Since air resources are limited, however, ISDM must take account of the fact that the date of availibility for airlift and for sealift may not be the same, both because of the possibility that the cargo must be moved to the POE and because it may take longer to prepare it for shipment by sea, e.g., in the case of containerization.

Every airlift of a package consumes airlift fleet utilization hours and increments the airlift clock. If the air clock is incremented to a new day, a new set of packages may become available and the airlift priority list may be changed. Figure 15 (based on [18]) gives a simplified schematic of the scheduling of packages with inconsistent availabilities and priorities, and shows how the scheduling rule that takes account of both may make maximum use of resources and accelerate delivery dates.

Figure 16 (based on [18]) illustrates the adjustment of schedules to take account of the constraints of numbers of aircraft that can take outsize and oversize cargos. Figure 17 (based on [18]) shows the further gain to be had in delivery capability by scheduling filler bulk cargo to available unused capacity in scheduled aircraft. As noted earlier, simulations do not guarantee optimum allocations, only feasible and "good" ones, but it is interesting to note that the utilization of ships in many runs is reported to be between 95 and 100 percent.

The problems of scheduling convoys, noted earlier, become more complex and more interesting when account is taken of attrition, priorities for replacement, and so on. As noted earlier, however, the attrition model (ISIM—Interactive Strategic Interdiction Model) has not been exercised sufficiently to validate it as one way of solving the problem.

PACKAGE	AVAILABILITY (AT BEGINNING OF DAY)	PRIORITY	DAYS TO MOVE
A	2	1	1
B	1	2	2
C	2	3	2
D	1	4	1

CASE I. SCHEDULED BY PRIORITY ONLY

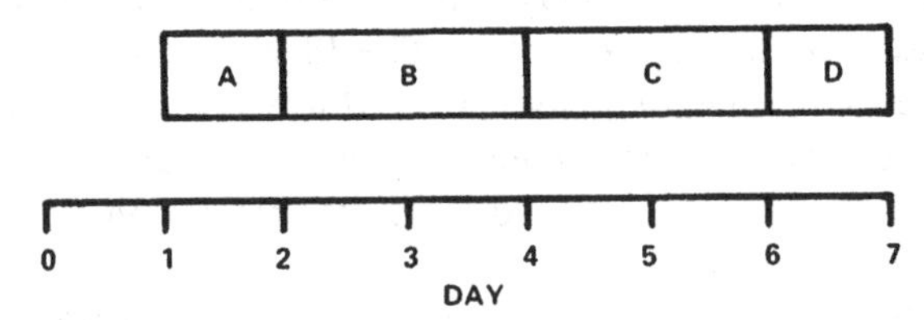

CASE II. SCHEDULED BY HIGHEST PRIORITY AVAILABLE

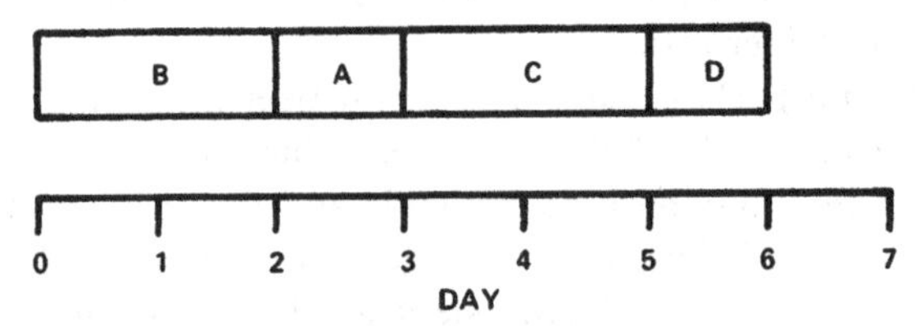

IN CASE I THE SCHEDULE IS STRICTLY BY PRIORITY. NO MOVEMENTS CAN OCCUR ON THE FIRST DAY BECAUSE THE HIGHEST PRIORITY PACKAGE IS NOT AVAILABLE TO MOVE. IN CASE II THE HIGHEST PRIORITY PACKAGE AVAILABLE TO MOVE IS SCHEDULED FIRST AND SUBSEQUENT PACKAGES BENEFIT BY ARRIVING EARLIER (DAY 6 VICE DAY 7).

FIGURE 15. Illustration of Two Priority Selection Rules

VALIDATION

It is not easy to say whether ISDM, the current version of the strategic logistics model, has been validated. The analysts involved feel that they have expended a great deal of effort in the attempt to validate. The consistency between the simulation results and earlier linear programming results has been taken as one measure of validity. The long usage and high degree of user acceptance is taken as another, indirect measure—though this criterion

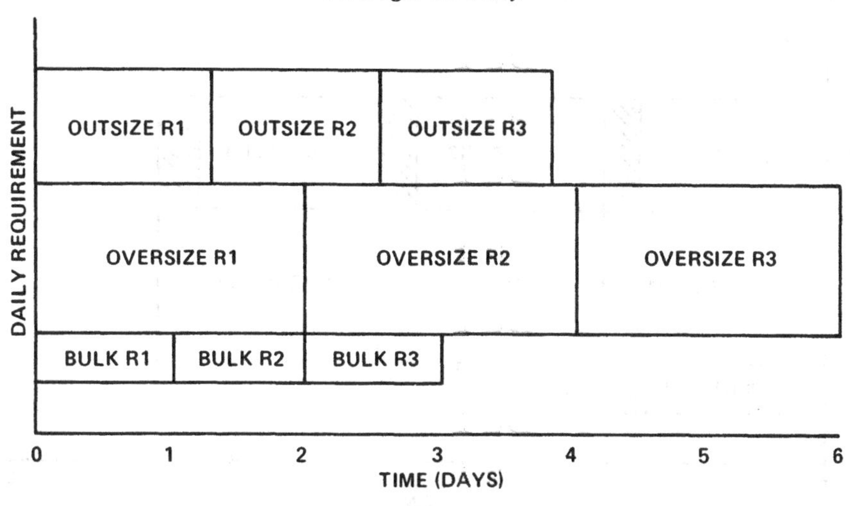

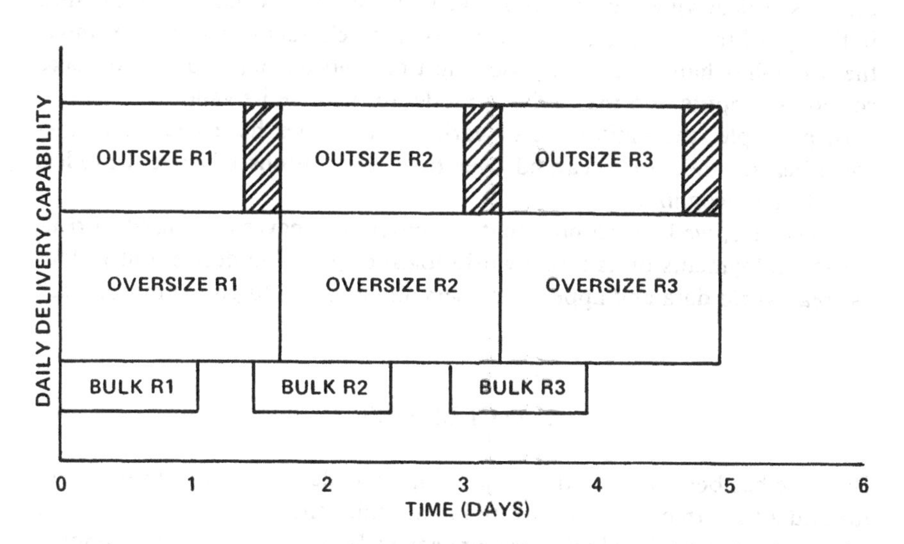

SHADED AREA REPRESENTS OUTSIZE-CAPABLE AIRCRAFT DELIVERING OVERSIZE CARGO.

FIGURE 16. Airlift Scheduling

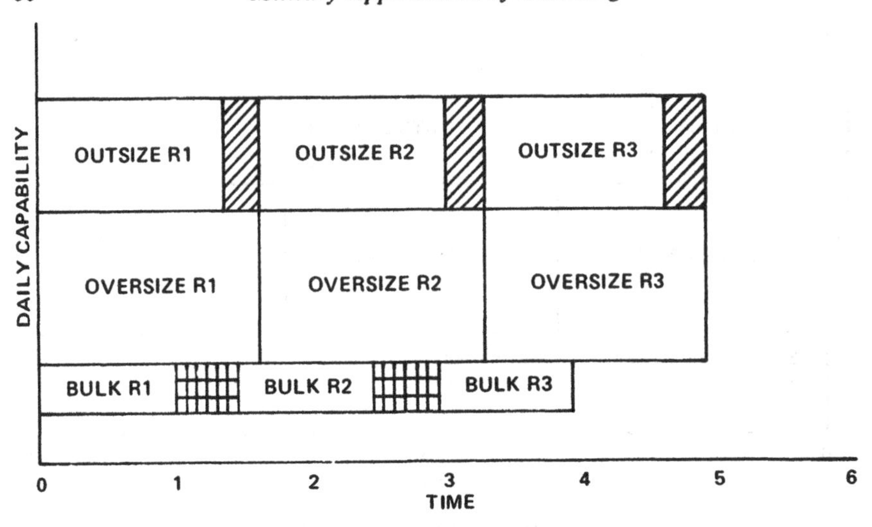

SHADED AREA REPRESENTS OUTSIZE-CAPABLE AIRCRAFT DELIVERING OVERSIZE CARGO.
CROSSHATCHED AREA REPRESENTS MOVEMENTS OF FILLER RESUPPLY.

FIGURE 17. Scheduling of Filler Cargo

can be self-deceiving if user acceptance is self-serving. Validity is sometimes in the eye of the beholder, but in this case the beholder—most importantly, the user—has had a very long viewing time and an opportunity to make continuing inputs into the evolving model to increase its validity.

In principle, the limitations of earlier approaches have been known—some beforehand, some learned over time. Limitations of current models seem to be recognized.

In practice, we have to note that the model has never been used in real-world deployments or as a real-world logistics planning device, but it does use real-world data and appears to users and analysts to give commonsense results.

CONCLUSIONS

This case has been presented in chronological order, covering almost a decade and a half, in order to illustrate in particular the evolutionary nature of modeling, in cases in which there appear to be high user/analyst rapport and communications. The case also illustrates the importance of technology as an engine of change in modeling, as in weaponry.

That the models have had good acceptance over this long period of time is evidenced by their continued use by OASD (PA&E) as well as DCA's Command and Control Technical Center (CCTC) in support of J-4, and in particular cases by both the Army and the Navy. One might especially cite the use of the model in 1977 calculations for PRM-10† and the Consolidated Guidance for FY1978.

How much influence the models have led is basically more difficult to assess. One would have to assess the influence of the users and of the larger studies in which the models were utilized—surely beyond the purview of this document. The models have been used primarily in force planning. We have noted that political factors have at times dominated analytical findings, such as when the Vietnam-reflex arguments, not analytical evaluations, defeated the FDL (Fast Deployment Logistics ship). Budget pressures and other influences have also clearly played large roles.

The strategic logistics models also illustrate the tremendous complexity of logistics, reflected in the size and complexity of the models and the computer requirements. Nevertheless, complexity is not the same thing as complicatedness. The author would have to conclude that the conceptual problems are not of the same order of magnitude as those introduced by the complications and less well-understood interactions of theater combat problems (Chapter VI).

Our study has a cutoff point, but not an ending. Perhaps the analysts, and surely the users, feel that they have just begun to fight. Perhaps the above statement about greater complexity than complication simply reflects the necessity of getting on with force planning without truly solving the problems of the measurements of effectiveness of logistics. To model tonnage delivered, or time of delivery, is to measure transportation, not warfighting, capability. Ben Franklin long ago cautioned us that "for want of a nail," etc., but he did not tell us how to measure the value of the nail, the shoe, and the horse, in the battle. The question is begged by the assignment of priorities to fighting units and to supplies. The analysts have very much in mind the importance of future work on the development of logistics models fitted to and adequate for inputting to the hierarchy of models necessary for the evaluation of theater combat, and so ultimately the logistics modelers will indeed share the conceptual problems of the theater combat modelers.

We have already noted the great gap of the omission of full modeling of the vulnerability/attrition problem. If the institutional problems are solved

---

† PRM is the acronym for Presidential Review Memorandum, the equivalent of the NSSM, or National Security Study Memorandum of the Nixon-Ford Administrations.

to allow this work to go forward, the conceptual and analytical problems will probably seem to multiply before our eyes.

Meanwhile, the nitty-gritty problems are always with us. There is a data base of fifteen to twenty thousand items which must be maintained, updated and probably expanded. There is also a need to extend the simulation to, or integrate it with, intratheater logistics modeling.

Questions

1. Why is logistics not a simple optimization problem?
2. How many kinds of uncertainties can you list that affect planning for the supply of overseas theaters of operation?
3. Discuss some or all of the uncertainties in Q. 2 and suggest ways of modeling them and likely sensitivities (which ones appear most likely to distort planning? to make outcomes unpredictable? to be decisive in outcomes?).
4. What different problems would you anticipate for an intratheater logistics model?
5. Discuss the scheduling "clock" in the ISDM. Can you think of alternative ways of handling the problem it models?

References

1. R. M. Nixon, *U.S. Foreign Policy for the 1970s: A New Strategy for Peace*, U.S. Government Printing Office, 1970.
2. Lee G. Wentling, Jr., George R. Fitzpatrick, Mary J. O'Brien, and Justin C. Whiton, "A Programming Model for the Design of Strategic-Deployment Systems," Research Analysis Corporation, RAC-TP-211, September 1966.
3. George R. Fitzpatrick, Jerome Bracken, Mary J. O'Brien, Lee G. Wentling, and Justin C. Whiton, "Programming the Procurement of Airlift and Sealift Forces: A Linear Programming Model for Analysis of the Least-Cost Mix of Strategic Deployment Systems," Research and Analysis Corporation, RAC-P-17, March 1966.
4. George R. Fitzpatrick and Justin C. Whiton, "A Programming Model for Determining the Least-Cost Mix of Air and Sealift Forces for Rapid Deployment," Research Analysis Corporation, RAC-P-19, July 1966.
5. Dr. David C. Dellinger, *et al.*, Dept. of Defense; Office, Assistant Secretary of Defense, Systems Analysis; "OASD Data Book (Working Data) for Strategic Mobility and Transportation Model," 1966.
6. P.M. Jenkins, M. J. O'Brien, and J. C. Whiton, "Description of the DOD Mobility and Transportation Resources Allocation (MATRA) Model," Final Report prepared for Office of the Secretary of Defense, Research Analysis Corporation, September 1967.
7. P. M. Jenkins Mary J. O'Brien, and Justin C. Whiton, "The Mobility and Transportation Resources Allocation (MATRA) Model," Research Analysis Corporation, RAC-P-41, June 1968.
8. *Gross Feasibility Estimator GFE-III User's Manual*, National Military Command System Support Center, CSM UM 37A-68, July 1968.
9. J. C. Whiton, M. J. O'Brien, P. M. Jenkins, and M. J. Dillon, "The DOD Strategic-Deployment Problem: Formulation and Techniques, 1967-1968," RAC-P-42, Research Analysis Corporation, October 1968.

10. J. C. Whiton, Mary J. O'Brien, P. M. Jenkins, and M. J. Dillon, "POSTURE: A Third-Generation Linear Programming Approach to the Strategic Deployment Problem," Final Report, October 1968.
11. P. E. Chesbrough and Leslie G. Lynch, "Capability Maximization Version of a POSTURE Linear Programming Formulation," Research Analysis Corporation, May 1969.
12. M. J. O'Brien, H. S. Weigel, and C. J. Keyfauver, "Description of the POSTURE Matrix Generator," Research Analysis Corporation, draft, November 1970.
13. J. E. Cremeans, W. C. Kilgore, W. O. Miller, and H. S. Weigel, "Analysis of J-4 Study Techniques," Research Analysis Corporation, draft, April 1971.
14. C. J. Keyfauver, "PRESHIP—A Preprocessor Program of the POSTURE System," Research Analysis Corporation, draft, May 1971.
15. W. C. Kilgore, "J-4 Study Analysis—A Strategic Mobility Modeling Procedure," Research Analysis Corporation, draft, November 1971.
16. C. J. Keyfauver, "User's Manual, QTYP," Research Analysis Corporation, draft, June 1972.
17. W. C. Kilgore, "A Mathematical Programming Technique for Strategic Mobility Force Capability Analyses," General Research Corporation, September 1972.
18. M. J. O'Brien, "POSTURE System Description and User's Manual," General Research Corporation, OAD-CR-5, June 1973.

Bibliography

C. J. Keyfauver, "The Interactive Strategic Deployment Model (ISDM)," General Research Corporation, August 1976.

C. J. Keyfauver, "The Interactive Strategic Deployment Model," General Research Corporation, August 1977 (draft).

CHAPTER IV

Modeling Strategic Bomber Penetration

INTRODUCTION

Modeling of U.S.-Soviet strategic nuclear exchanges started in a serious way in the early 1960s. The models have always had a certain air of unreality. This unreality stems, in the first instance, from the fact that there have been no nuclear wars (the one-sided demonstrations in Hiroshima and Nagasaki notwithstanding) and empirical data have not been available. This fact has freed analysts to make their own assumptions, "realistic" or not—realism being undemonstrable. A strong logical case can be made that the majority of assumptions popular in the defense community have been unrealistic in several respects: (1) in using population fatalities as the primary measure of effectiveness; (2) in adding industrial capacity (using industrial floor space as a proxy variable for manufacturing value added, because floor space can be estimated from aerial or satellite photography)—industrial capacity being highly correlated with population distribution, so that it didn't matter anyway in projecting urban-industrial attacks; (3) in postulating only massive, one-time exchanges. There has, indeed, been increasing sophistication and variation of these assumptions in recent years, as there has been deepening concern with the implications of parity or worse. A recent study has, in fact, listed forty-one possible different measures of the relative power of the U.S. and Soviet strategic nuclear forces.[1]

We will come to strategic exchange models in Chapter VII. But, except for counterforce, they do not illustrate offense/defense interaction (ABM being essentially ruled out, and civil defense just beginning to be seriously analyzed, and at that only parametrically and still largely in terms of population defense only). More interesting, therefore, is the strategic bomber attack case, since the United States faces air defenses and there are empirical data on bombers versus air defenses. We have chosen to use as a case history a model of bomber penetration of the Soviet Union. Note that an exchange model for bombers is of no particular current interest, since the U.S. has no significant air defenses. Should this situation change, the model we will discuss could be adapted to playing both sides of the game. More importantly, it can provide useful measures of bomber capability as a component of overall strategic exchanges.

THE ADVANCED PENETRATION MODEL (APM)

The Advanced Penetration Model, hereinafter APM, has at time of writing a ten-year history. In response to a number of perceived analytical and planning needs in the Air Force, Air Force Studies and Analyses in late 1968 recommended the development of an authoritative strategic bomber penetration model. The first months of 1969 saw the conceptualization by a committee of Air Staff and SAC officers of the requirements and approach for such a model. In September 1969, a contract was let to Boeing Computer Services for model development. No less than sixty personnel (twenty analysts and forty programmers) were initially committed to the task—a measure of both the complexity of the task and the importance attached to it by the Air Force. It was some three years before the first major operational application of the APM was undertaken in the Joint Strategic Bomber Study (JSBS).

The JSBS, involving some ten full-time officers, under the jurisdiction of the Joint Chiefs of Staff, compared a number of mixes of equal cost forces of B-52s, FB-111s, the B-1 then in development, and the FB-111H, a hypothetical modification of the FB-111 designed primarily to increase its capacity. These bombers carried alternative mixes of weapons, including gravity bombs, SRAMs (Short Range Attack Missiles) and ALCMs (Air-Launched Cruise Missiles). The results of the JSBS were discussed in Senate testimony on March 19, 1975, and an unclassified summary was read into the record in 1976.[2]

We will return to the JSBS and the controversies over its usefulness after we have described the model. We will also note the very considerable number of other applications the model has had since that time. The simulation approach of APM permitted the use of a number of different measures of effectiveness, and the original concept was that it should be highly flexible. Among the measures that could be developed were the following:

number of weapons delivered; number of targets destroyed; total value of targets destroyed (on the user's value scale); depth of penetration of bombers; numbers of bombers surviving and recovered; cost of a given delivery capability; fuel consumption in alternative attacks; and so on. Attacks could be massive and by many routes, or limited and concentrated.

The APM computing efficiency has improved by an order of magnitude as it has gone from the IBM-360 series to the -370 series and then to the -3032. The later has a 3-megabite capacity and virtual memory. It is more powerful (larger core capacity and faster CPU) than the Honeywell 6180 MULTICS, though the latter is sometimes used to run components. In addition, the 3032 compiler has increased data management efficiency two- to threefold. The current version is in the process of being transferred to SAC. It is of interest that, in contrast to, say, the series of logistics models described in Chapter III, the basic concept of the APM has not changed during the first decade of its life, although the model has been continuously upgraded by refinements and expansions of its capabilities (including some details that add more to verisimilitude than to analytical usefulness).

SIMULATING THE MISSION OF THE STRATEGIC BOMBER FORCE

The ambitious plan of APM is no less than the simulation, by individual bomber (and tanker), of the entire strategic mission of the entire bomber force, or any part thereof, from takeoff to landing at a recovery base. Figure 18 (based on [3]) shows the typical flight path of a penetrating bomber that is modeled in the APM. The mission starts with the takeoff of the bombers and tankers, possibly under attack by Submarine Launched Ballistic Missiles (SLBMs), (which may be launched closer to shore than implied by the notional diagram and may use depressed trajectories to achieve minimum warning time). Surviving bombers are refueled by surviving tankers and then run the gauntlet (or seek to avoid) enemy defenses, from offshore AWACS (Airborne Warning And Control System)-type systems and CAP (Combat Air Patrol) interceptors to local SAM defenses, make their bombing runs, and then exit to a recovery base. While the model focuses on the penetrating bombers, a cruise missile can be treated as the penetrator. That is, the profile can be that of a stand-off cruise missile carrier launching cruise missiles at, say, five hundred and more miles offshore. Interaction with other events such as defense suppression and cross-targeting with other offensive systems can be handled by exogenous inputs that affect the defense weapon and target arrays. The possibility of recycling of

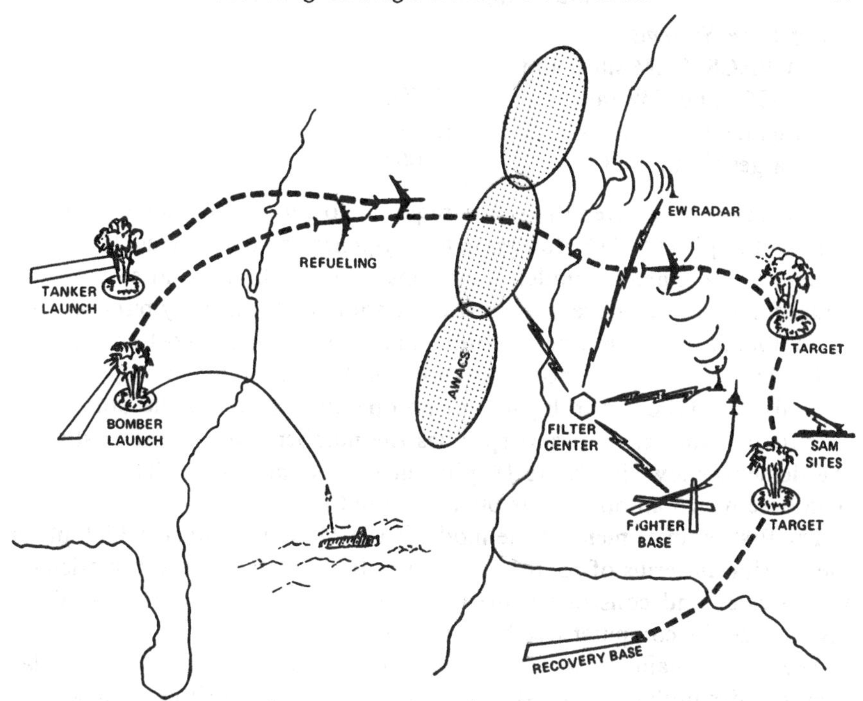

FIGURE 18. Representative Bomber Profile

recovered bombers is also exogenous to the model. Assumptions could be made about the recovery and survival of bombers to constitute the attacking force in a new APM run.[3]

The real significance of the APM is found in the fact that all of these complex offense/defense interactions are modeled for many bombers and many defense units simultaneously, that is, many-on-many, or force-on-force. Thus, the complex realities of defense saturation, command-and-control limitations, and weapon assignment doctrines can be captured in the modeling process. The forces can be very large. The present capacity of the APM is sufficient for the following numbers:

Offensive Systems	
Bombers	650
Tankers	650
Cruise Missiles	5,000
SRAMs and Bombs	5,000
Precursor Vehicles	5,000

Defensive Systems

AWACS, SAM sites, and GCI and EW radars	3,500
Fighter Interceptors	(no limit)
Target Complexes	4,000

This capacity is greater than that required to model the employment of present and planned U.S. forces on a fully-generated alert status. However, these are not inherent model limitations. Should the need arise to study larger forces in the future, the question is not one of feasibility but only one of additional data bank, programming, and computer run cost and time.

The basic logic of the model, as shown in Figure 19 (based on [3]), is essentially that followed in laying out an operation plan. At the outset, the user defines the scenario and specifies the numbers and characteristics of the elements shown in Figure 18, plus the bomber armament. These are the data base which an operations planner would use.

The first main element of the model is the Mission Planner, which plans the mission in terms of routes and targets for each bomber automatically, within rules and constraints specified by the user—a tremendous labor-saver, once the computer overhead is paid.

The second main element is the simulation of the air battle. The simulation includes both deterministic and probabilistic events. It uses the Monte

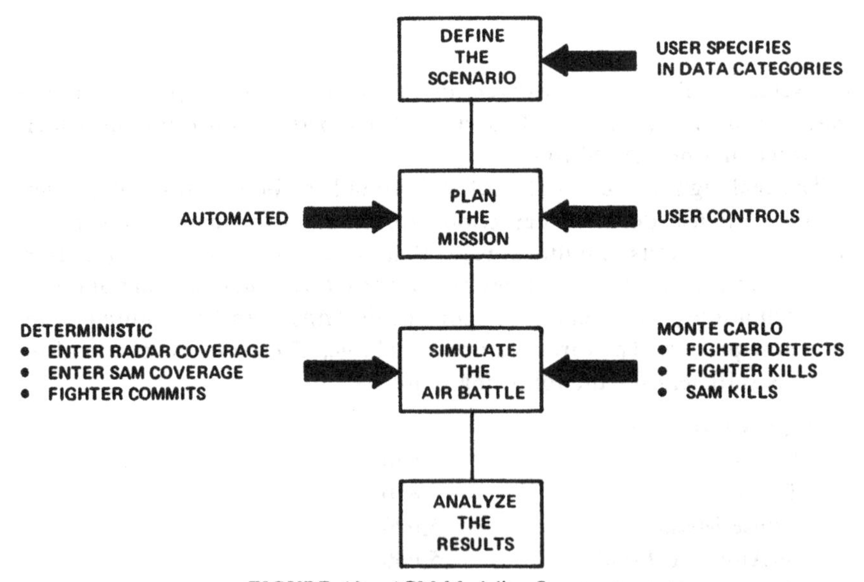

FIGURE 19. APM Modeling Sequence

Carlo technique for the treatment of random events such as detection, acquisition and kill for both fighter and missile interceptors.

All of the results of the air battle simulations are stored as output event notices (OEN). The file of these OENs is the data base for air battle reports that can be specified by the user, in accordance with his requirements. Thus, a user might be interested in target destruction (DE—damage expectancy), by category or area; weapons delivered, by category, bombers surviving, possibly by entry point, exit route, or other grouping; or the recall of the more detailed events to aid in the analysis of alternative mission planning rules.

There are at present some seventy-two data categories for geography, offensive and defensive weapon system numbers and characteristics, and so on, that the user can specify, with game controls for their manipulation. This number tends to grow, as additional variables are added. A recent example is the addition of data defining the global wind patterns at different flight altitudes. Game controls, or rules and constraints, can be added by the user with respect to each data category.

The language for APM/user interface is QUERY. The user can query the APM in English commands as to the status of any data element and thus can specify change for a specific item or category. At present there are over 200 thousand data elements in the seventy-two categories.

THE MISSION PLANNER

As noted earlier, the Mission Planner automates the complex task of preattack planning. The user can specify rules and constraints for the use of each category in the data bank. The Mission Planner then generates detailed plans for each sortie, including routing, refueling, target allocation, and recovery. As shown in Figure 20 (based on [3]), the Mission Planner consists of a series of modules, the output of one serving as the input to the next. The modularization of the model permits the use of each model separately, where desired. For example, the refuel module can be used for independent investigation of large force refueling and tanker requirements. The content and function of each module is discussed below.

Grid

The geography of the potential enemy country is defined by latitude and longitude in a grid of five degree by five degree areas. All identified targets are located on this grid, and each cell containing one or more targets or de-

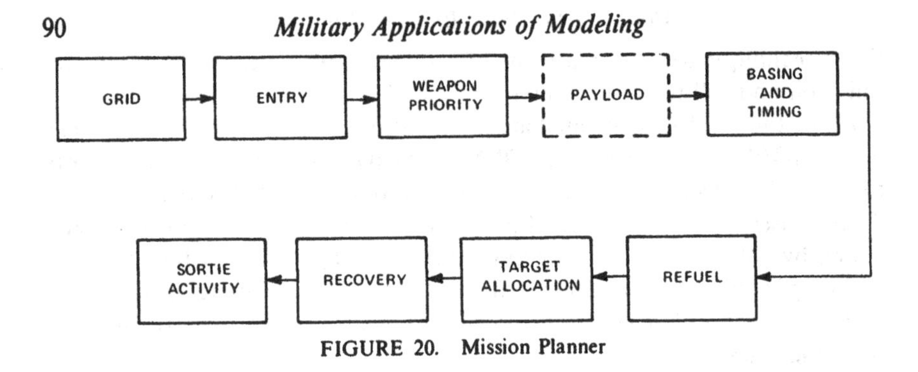

FIGURE 20. Mission Planner

fense elements is given an ID number. The target data base so organized provides an input to both the selection of entry points and the allocation of targets.

Entry

Potential entry points, which can be user-selected, are defined in terms of the distances to the nearest target on the grid. Distances to defenses are also specified.

The model does not, however, restrict allocation of given targets to bombers using one entry point only. Since many in-country flight paths are very long, crossovers may often prove advantageous.

Weapon Priority

The weapon priority module, the logic of which is shown in the example in Figure 21 (based on [3]), allows the user to set priorities for the allocation of weapon type to targets. In the illustration, the weapon mix includes SRAM, cruise missiles and gravity bombs. If the target is defended, the user can specify that gravity bombs not be used. If an undefended target is hard, the use of a gravity bomb may be preferred because it is cheaper than a cruise missile or SRAM. All that this module does is to specify the priority for using available weapons against given target types. On a given sortie, the load of a preferred weapon may have already been expended by the time a given target is reached. Note that the question of defense suppression with bomber weapons (or ballistic missiles) is deferred to the target allocation stage.

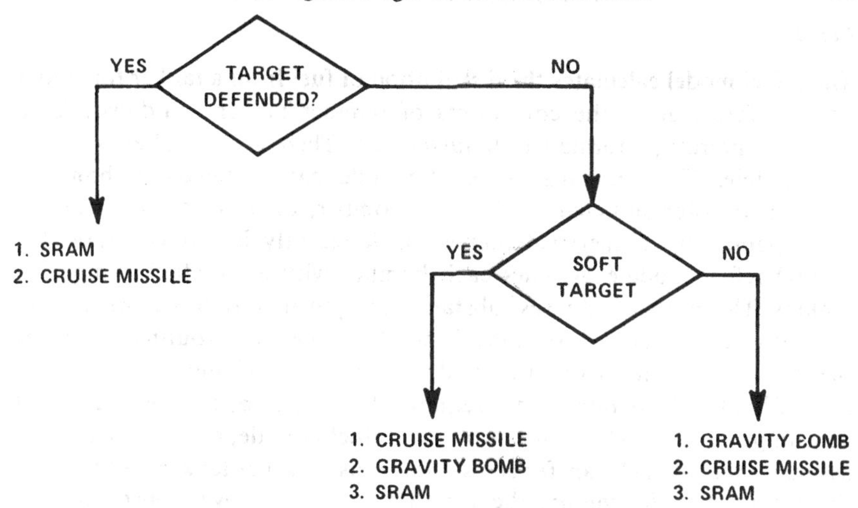

FIGURE 21. Weapon-Priority: User Can Prioritize Weapons

Payload

The payload module automatically loads each bomber with the type and number of weapons required for the targets associated with the entry point of that bomber. The module is shown in dashed outline because it is frequently not used; the algorithm needs improvement, and direct user input is sometimes simpler.

Basing and Timing

In defining his scenario, the user may specify entry points to a target country, the launch bases for given entry points, and detour points to avoid overflying specified areas. The user may also specify a desired arrival time at entry point, in order to control the separation between bombers. The basing and timing module computes the distance from launch bases to entry points and determines the arrival time of each bomber at entry, taking account of user-specified constraints.

If the user does not specify launch points for each point of entry, an associated LP routine called "optimum basing and timing" will optimize launch-to-entry-point combinations to minimize the total distance flown, by bomber-type, for the fleet.

Refuel

The refuel model calculates the distribution of fuel from a tanker force to a bomber force, given the constraints of bomber/tanker bed-down, force size, and aircraft performance characteristics. There are two alternative refuel modules. The original zone refuel module matches tankers to bombers so as to transfer sufficient fuel to the bomber, to prevent arrival at the entry point with a specified fuel level. A recently introduced ratio (or buddy) refuel module matches each bomber with a specified number of tankers. This module achieves substantially equivalent results more quickly and easily, at the sacrifice of some detail. The zone refuel routine minimizes the number of tankers required and takes account of limitations such as fuel offload/onload rates, time required for hook-up, and refueling fuel consumption rates. The ratio, or buddy, refuel module, on the other hand, assumes that the fuel transfer is instantaneous and that tanker and bomber flight paths to refueling are the same (ignoring bed-down constraints). It gives a rapid approximation to tanker force requirements (e.g., one tanker per bomber or one-half tanker per bomber, or it can calculate the performance achievable with a given tanker/bomber ratio, in terms of fuel availability for the bombers at point of entry).

Target Allocation

Crucial to mission planning is the allocation of targets to bombers. The task is detailed and tedious, but fully automated. The user, however, can control the allocation in several dimensions. At the outset, each target is assigned a score, calculated by dividing a specified power of its assigned value by a specified power of its distance from the nearest point of entry for penetrating bombers or cruise missiles. Clearly, higher-value targets should be preferred, as should closer, or less distant, targets. The assignment of values is, unfortunately, necessarily arbitrary, a problem that plagues all targeting schemes. Within a given class, there is usually some simple measure for the ordering of targets by relative value. A furnace producing a million tons of steel a year is clearly twice as valuable as one producing five-hundred-thousand tons a year. Inaccuracies in estimating capacity can perhaps be ignored, and differences in the quality of steel perhaps adjusted for. But between classes of targets there are seldom common denominators permitting objective ordering. On the other hand, the distance from point of entry is measurable and its significance presumably proportional to the estimated probability of the survival of the penetrator to that point. Survival against area defenses is, in general, proportional to time of exposure, or distance flown. (This may be modified by factors derived from experience on the value of massing at the point of entry in order to saturate defenses.) Given the assigned values and geographic data, the Mission Planner calculates the target score, equal to value^x/distance^y. (In the model, bomber losses turn out to approximate an exponential function of penetration distance, with an exponent of somewhat less than one.) It then has the first bomber entering at a given entry point search over a specified range and angle in azimuth for the highest-score target on its path. The next bomber searches for the highest-score remaining target, and so on in time-ordered sequence. Since the probability of destruction of any given target by a single weapon is less than one, the value of a target is degraded when it is attacked, but its adjusted score remains in the target array. Thus, very-high-value targets may be attacked more than once. The user controls all of these variables that constitute the rules for the target allocator, including the values of the exponents x and y that govern the relative weights of target values and distance, or risk involved. The earlier module for weapon priority has already inputted the user's preference on weapon priorities; particular classes of targets can be given preference in the target allocator, and specific targets can be assigned if the user so chooses.

Up to this point, the target allocator has assumed a great-circle flight path from point of entry to first target and between each pair of targets. With the targets now allocated, a flight path selection routine seeks a path to each target that avoids or minimizes exposure to local defenses and to weapons effects.

If GCI radars and SAM sites cannot be entirely avoided, the model introduces user-specified ECM to degrade the effective radius of the defenses. If this still does not clear a feasible flight path, the model will dictate the launching of a SRAM for defense suppression, if the user-imposed game controls so permit. Again, a damage expectancy (DE) against the defense site is assumed. Additional SRAMs can be assigned, and when the DE reaches a specified threshold the defense site is assumed to be killed and need no longer be avoided, given that specified rules for the avoidance of weapon detonation areas are observed.

As in the target selection phase, the automation of flight path planning saves a great deal of time, although considerable control is still in the hands of the user. As already noted, the user can specify ECM use and effectiveness. In addition, he can specify the maximum deviation from a great circle routing, maximum number of turns, effective radius of the defense site, and avoidance priority (e.g., which of two SAM types is a greater threat and therefore should be avoided, if exposure to one or the other is inevitable).

Note that all SAM sites are treated as fixed. This may be a serious weakness in the model. While mobile defenses cannot be taken into account in mission planning, a bomber may be able to detect such a site when it turns on its radar and may be able to take evasive or suppressive action. A cruise missile—at least in any version likely to be deployed in the early to mid-eighties—can do neither. Thus, the model may seriously overestimate the penetrativity of cruise missiles relative to manned bombers. (The other major vulnerability of cruise missiles, that of their carriers off-shore, before missile launch, can be handled in the APM by user adjustment of the location of the AWACS, or SUAWACS, and of interceptor kill probabilities.)

Recovery

Finally, the mission plan includes a routine that flies the bomber from the exit point to the closest allowable recovery base that is suitable for the given bomber type and the capacity of which has not already been filled. If no such base is within reach, a bomber will be recovered at a base that fails to meet the criteria.

Sortie Activity

The sortie activity module is not a mission planning step, but rather an output data management function. It sorts the event list that has been generated by time and distributes the weapons that have been allocated to target complexes in the grid to DGZs within each complex according to a pre-established DGZ ranking. The final product of the Mission Planner is a penetrator event list covering, for each bomber, the time of takeoff, start of cruise at altitude, first and possibly second refueling, jettisonning of external tanks, entry, altitude and speed changes, turns, weapon releases, exit and recovery. This event list can be used as input to the APM air battle simulation or other simulation models.

The APM is planned on a modular basis, so that separate parts can be used for special purposes or, in some cases, be omitted. The Mission Planner can be used as a tool in itself, as can its modules. For example, the refuel module could be used separately for the calculation of force refueling requirements, and perhaps for the comparison of some characteristics of alternative tankers.

AIR BATTLE SIMULATION

At this point, the user and the Mission Planner have done their homework. Now the battle can begin, or at least be simulated. Figure 22 (based on [3])

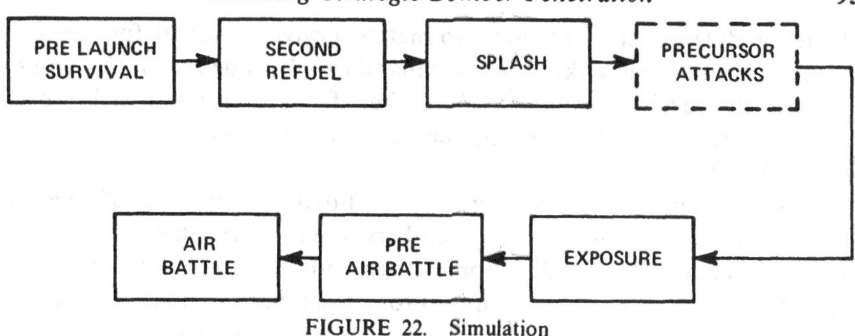

FIGURE 22. Simulation

shows the steps in the APM simulation. The first six boxes represent preprocessors through which the event list, or stream, must pass before the actual simulation of the air battle. Again, the structure is modular, and components can be used singly. The prelaunch survival module, for example, might be used for comparisons of different aircraft or for the study of alternative basing concepts for given aircraft. As above, the module will be briefly described, largely on the basis of Reference 3.

Prelaunch Survival

Figure 18 implied that the war or the battle started with an attack on the bomber bases by SLBMs (or ICBMs, cruise missiles, or even sabotage). In other scenarios, this module might be omitted. Given a specified attack, the model takes account of user-specified alert status of forces, characteristics of the attack, and the warning received by the aircraft; the model determines by the Monte Carlo method which bombers and which tankers survive and take off. (Since a pattern attack around an air base is generally considered the most severe threat, this step may be envisioned as including the survivability through flyout.) The prelaunch survivability (P_{LS}) of an aircraft is a function of its takeoff time, flyout speed, and hardness to nuclear effects. P_{LS} will therefore vary by aircraft type. The variation among bomber bases is computed by offline routines and inputted into the Prelaunch Survival module.

Second Refuel

Second Refuel matches surviving bombers with surviving tankers. As noted above, a zone refuel or ratio refuel module can be employed.

If there has been a bomber base attack, it cannot be expected that surviving bombers and tankers will match up as initially planned. The Second

Refuel module therefore attempts to match surviving bomber fuel requirements with surviving tanker offload capabilities. In zone refuel, this will be done for the surviving bombers and tankers from the list assigned to each zone; in ratio, or buddy, refuel, the surviving bombers and tankers are matched up for each base.

In either event, if the surviving tanker offload capacity is insufficient to meet the demand of the surviving bombers, the routine will offload a user-specified fraction of individual bomber demands, i.e., each bomber will get a fixed percentage of what it "requires for its planned mission." Depending on the refueling percentage selected, some surviving bombers may go unrefueled. The user may assign priorities for the refueling of bombers.

Splash

This vividly, not to say starkly, named module computes the point in its planned flight path at which each incompletely refueled bomber will run out of fuel.

Precursor

Precursor allows the user to assign ballistic missiles to degrade defenses before the bombers arrive. (Again, the dashed outline is to indicate the step may be omitted.) The user specifies the precursor weapons, CEPs, yields, and aimpoints. The module calculates weapons effects and determines which defenses are killed. Precursor takes no credit for collateral damage to the target base, but this is done in the Nuclear Effects Processor (see below, p. 148).

Exposure

The planned flight path of each penetrator is preprocessed to calculate its exposure to defenses. This generally starts with the AWACS, for which the synchronization of the moving radar envelope is calculated, plus postulated CAPs. Line-of-sight and instrumented range exposures to ground-based radars (EW, GCI, and SAM) are also calculated. These calculations assume a smooth earth with a 4/3 radius to account for refraction antenna heights, masking and elevation level limits. Penetrator altitude is also taken into account.

Pre-Air Battle

The final preprocessing generates no new events but takes the events that are the outputs of the previous modules and edits and merges them. No further user control is feasible at this point. The Pre-Air Battle processor is a fully automated system for the time-ordering of all bomber events from point of entry to the defended air space to point of exit and the tabulation of the identification, characteristics and capabilities of all elements, or players, both offensive and defensive.

Air Battle Simulation

At last, the air battle—the interaction of bombers and defenses over the enemy's defended air space—can really begin. Air Battle Simulation is a time-sequence process of the events planned in the Mission Planner and modified in the air battle preprocessors. Each event will generate the subsequent event, on a deterministic or probabilistic basis, as appropriate. The process starts with the entry of the first bomber into the radar coverage. It will continue until the exit of the last surviving bomber, unless it is interrupted by the user.

There are seven air battle event processors:

1) Command and Control
2) Radar
3) AWACS
4) Interceptor
5) SAM
6) Nuclear Effects
7) Penetrator

Each event is identified with one of these processors by the first letter of its "unique event identifier." For example, when the first bomber entered the radar coverage, it may have generated an "R006" event; this event is processed by the Radar Processor. It generates a "C005" event, fighter-requested, in the Command and Control processor.

Command and Control Processor

The command and control net is of particular interest. Figure 23 (based on [3]) illustrates this net, which controls only area defenses, or interceptors. SAM sites are assumed to operate autonomously. A C^2 processor accumulates threat information, evaluates it, and assigns interceptors. The user can set saturation limits on the elements of the C^2 net, such as the maximum

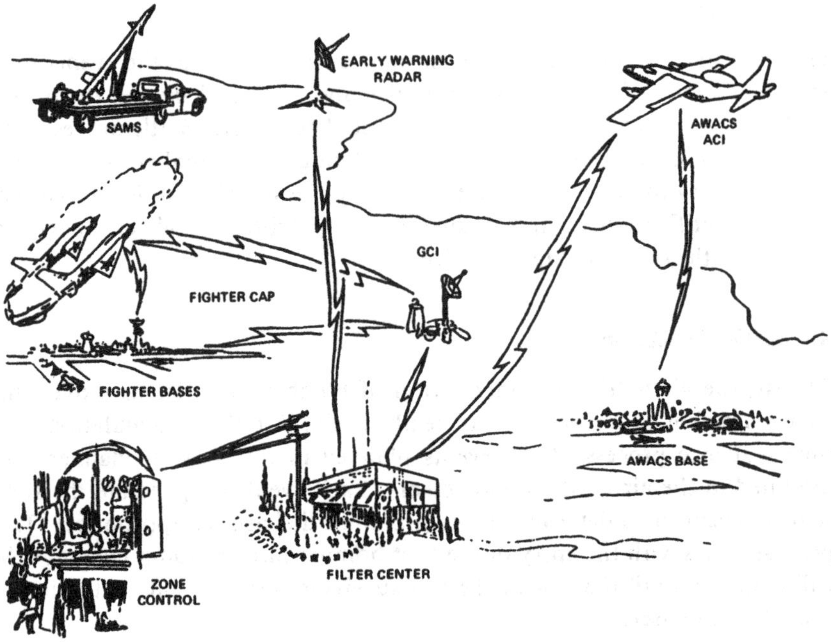

FIGURE 23. Command and Control Net

number of interceptors that can be controlled by a filter center or GCI (Ground Controlled Intercept) radar site. When a command element is saturated, new threats go into a queue and are acted upon as the element becomes unsaturated. If a filter center is killed, breaking the C^2 net, the remaining elements of the net continue to function autonomously.

Radar Processors

The simulation uses a Radar Processor for each of the radar categories: airborne (AWACS and fighter) and ground-based (GCI and SAM). The Radar Processors record the radar characteristics (instrumented range and line-of-sight range), radar scope condition status, reflecting penetrator ECM use, and penetrator detection.

Penetrator detection will vary with the radar cross-section (RCS) of the penetrator. RCS is a function of the aircraft type and of the angle of view. For each penetrator type, two RCS values are entered in the model: a low RCS, the average value for nose-on to a sixty-degree aspect angle; and a high RCS, the average of sixty degrees from beam to the stern view. The Radar Processor computes the point in time of entry to the radar coverage, detection probability, and, if the penetrator was not detected at entry, the point at which the RCS crosses over from low to high value, in order to assess a second detection probability. The Radar Processors also take account of ECM. When a bomber detects that it is entering radar coverage, it attempts to jam the radar with noise, causing a strobe to appear on the enemy radar scope and denying him range information. From the user-specified jammer power, bomber RCS, and defense radar characteristics, the processor computes where and if burnthrough will occur. The user can also specify a limit on the number of simultaneous strobes that a radar can process. When the threshold is exceeded, the radar no longer performs as a censor or invectoring interceptor. New penetrators entering the coverage of the strobe-saturated radar will not be detected, but if they are jamming their strobes will be added to the strobe counter for that radar.

Interceptor Processor

A penetrator detection event generates a fighter request by the Interceptor (aircraft) Processor. Available fighters belonging to a given fighter net are polled and the aircraft that can make the fastest intercept is assigned. The fighters are vectored to radar range, where the probability of detection and conversion is Monte Carloed. If the fighter passes the P_{DC}† checked, it advances to the end-game (or missile-firing) position, where a probability of kill check is made. If the P_K check is passed, a penetrator kill event is posted, the fighter returns to its base, and the penetrator event chain is terminated.

The Interceptor Processor can also handle the case in which the bomber has ECM turned on, causing the radar to show a strobe which denies it range information. An interceptor can "ride the strobe" toward the bomber, but Command and Control will not know whether P_{DC} turned out to be 1.0 or 0.0, i.e., whether the P_{DC} check was successful.

User inputs can degrade P_{DC} for strobe-ride, ECM, out-of-radar coverage, and penetrator dash conditions, and can also degrade the fighter's P_K for ECM and IRCM (infrared countermeasures). These degrades can be varied by both penetrator and fighter type.

SAM Processor

The modeling of the SAM-bomber duel is fairly straightforward. The processor takes account of the time at which the penetrator comes within the

---

†P_{DC} = probability of detection and conversion.

instrumented range of the SAM site and the time and probability of its detection. It then calculates the times for launch and flyout, the point of intercept, and the probability of kill. The P_K reflects several factors. First, for each SAM type a grid is stored in the processor. This grid is ten by ten, and each cell has an associated P_K. Zero values represent dead zones, and missiles are not launched for intercept in these cells. The P_Ks can be degraded by bomber ECMs and expendables (self-defense air-to-air missiles). These values are inputted by penetrator and SAM type. Actual kills are then Monte Carloed. If there is no kill, the processor will calculate whether there is time for another launch.

The possibility of saturation of the SAM site is allowed for, and a first-intercept-possible/first-launch rule is used for launches against a queue of bombers.

SAM sites are assumed to be points, that is, the missile batteries and radars are collocated. The sites are also assumed to be independent, although limited communications between them can be modeled by the user control of SAM system and radar reaction times.

Nuclear Effects Processor

The point of detonation of a weapon is determined by drawing a uniformly distributed random number against the azimuth of weapon impact and a normally distributed number in range, to simulate CEP. Damage from the detonation is calculated for each installation exposed (with hardness specified), and the value of that target is reduced accordingly. Damage assessment can be read out by target class and type.

Results

In addition to the above damage assessment, the product of the simulation is the record of the penetrator event list. There can be many events for each penetrator, starting with his entry into the AWACS line of site, his detection, coming under attack by a fighter plane, turning, switching on ECM, etc. The total number of events for a large force can approach ten thousand—say, an average of fifty events for each of two-hundred penetrators (some of which may be intercepted by AWACS-controlled CAPs, others of which may have fulfilled their planned missions, launching all their weapons and reaching their recovery base). Each event is recorded in the form of an output event notice. An OEN is a 45-word record. This record includes the type and time of events, the penetrator event and type, and the parent event that caused it to be generated, so that the event sequence can always be reconstructed. Additional information may include geography, interceptor of SAM ID, command and control agency ID, and P_{DC} and P_K.

The OEN provides basic data for air battle reports. Some of these reports are built in, or "canned". The user can access the OEN file with simple FORTRAN routines. Patently, the amount of data available from one run is tremendous. Forty-five words for each of one hundred thousand events would be four-and-one-half million words. The user must be cautious in requesting information, lest he drown in data.

ANALYTIC CAPABILITY

The wealth of data inputted to and outputted by the APM is both its strength and its weakness. The interaction of strategic offense and defense is modeled in great detail. The detail is such that an APM run for a large force, say two hundred alert, surviving penetrators, could require over twenty hours of dedicated time on the early IBM-360s, to progress from the primary data bank through the last air battle report, if everything worked right the first time! Sensitivity tests by means of repeated runs were inordinately expensive, although the modular construction of the model permitted the experienced and skillful user to modify the inputs to test numerous excursions without paying the price of entire APM runs for each.

The current IBM-3032, on the other hand, runs in minutes, and sensitivity tests can be afforded. Note also that although the model is probabilistic it does not require repeated runs for a given set of inputs. With 40,000 to 50,000 "rolls of the dice" in every full run, the law of large numbers ensures stable results.

The cost of development of the APM has been very high, as indicated earlier. The cost of ownership—maintenance and inputting of the data base and maintenance of the computer routines—is also large. The payoff is the data output for elaborate statistical analysis of offense/defense interactions (see below).

VALIDATION

As in most cases large models, validation of the APM turns out to be virtually impossible at a level that will satisfy critics as well as users. In a case in which the implications of the model for decisions of national import are so direct and so strong, the difference between users and critics tends in fact to be blurred. This may sound strange, but consider the fact that the direct users are the Air Force, and sometimes JCS and OSD, and the decision-makers they serve. But when decisions involve the types and numbers of bombers and air-launched weapons, not just their technical characteristics and tactics for their employment, DoD decision-making is subordinate in practice as well as theory to Executive Office and Congressional decision-making. Thus, even if AF or DoD users are satisfied as to the validity of the model, higher decision-makers may also be critics.

It is, the author believes, fair to say that it is beyond the power of an individual to understand fully the interactions that take place in a model as complex as the APM. Nevertheless, one can understand the pieces, or steps, taken one at a time, as well as the logic that combines them. In these terms, the Air Force appears to believe the APM has conceptual, or face, validity. And the data base, including both intelligence and technical characteristics, draws on vast Air Force and intelligence community efforts at empirical and experimental validation.

But the comments of the critics bear examination. One of the principal points of attack has been the handling of ECM, patently a key element in estimating bomber penetration.

ECM effectiveness is actually estimated off-line, in a model called COLLIDE, and inputted to the APM. COLLIDE is a model of the kinematics of bomber-interceptor interaction. The basic model includes procedures for calculating the probability of a maneuvering bomber entering the coverage of a GCI or AWACS radar and of its leaving that coverage before an interceptor arrives at a point close enough to attempt an intercept—i.e., an estimate of P_D, the probability of detection. To this has been added a set of parameters to calculate the probability of the interceptor radar acquiring the bomber, "converting" or taking over from the vectoring radar, to give the overall acquisition probability, P_{DC}. The model takes account of an eliptical envelope (with parameters A and B) around the bomber, defining the radar visibility of the bomber, the long axis of the ellipse being perpendicular to the longitudinal axis of the aircraft. (A refinement, using separate ellipses for the nose and tail of the bomber, is also available.) The interaction of these parameters with the angle of approach of the interceptor to the bomber (or, more precisely, to a point ahead of the bomber) is modeled. A weighted average of the possible angles of approach can be calculated for a random search and for a vectored search that reflects command and control capabilities and effects. All of these factors can be calculated on the basis of maximum theoretical detection ranges and of estimated burnthrough ranges when the relative powers of bomber ECM and defense radars are taken into account.

This is a highly simplified model of the complex ECM game. It assumes that a radar either detects or it does not, ignoring "partial detection" of the bomber by observing its ECM radiation. It assumes that if an airborne intercept (AI) radar acquires a bomber, there are no time delays or lock-on breaks. It does not take account of range and angular uncertainties that ECM may create, after burnthrough. It assumes that the interceptor pilot makes no errors by launching an AAM when he is outside the launch envelope or not launching when he is within the envelope.

An evaluation of the validity of this simplified model of ECM effects is difficult, and elements of subjectivity are involved. Nevertheless, there are some biases against the bomber, and it is hard to fault the Air Force for pro-bomber bias in this case. It is also difficult to see a bias in effects on the evaluation of different bombers.

The APM cumulates one-on-one engagements to reach many-on-many, or force-on-force. No one knows how to model ECM for force-on-force, but it is clear that in these circumstances the total effect of ECM—introducing noise and confusion, and enhancing the effects of saturation—will be greater than the sum of one-on-one estimates.

A second criticism is somewhat more theoretical. It is pointed out that one cannot tell how far the model deviates from the optimum in allocating bombers to targets. There is no reason to believe, however, that the results are not reasonably near optimum. In any event, the approach is conservative with respect to bomber performance.

Two specific criticisms bear on the question of SAM site effectiveness. First, the Command and Control Processor internets the C^2 of interceptors but not SAM sites. But Soviet practice is to do both, potentially greatly increasing the target acquisition capability of the SAMs. Second, the SAM sites are fixed. The Soviets are increasingly moving toward the use of mobile SAMs. These may still be avoidable or suppressible by bombers but not at all by preprogrammed cruise missiles. Since the Soviets tend to deploy mobile SAMs in relatively restricted areas for the protection of high-value strategic targets, it should be possible to modify the model to include probability distributions, possibly time-varying during the battle, for SAM sites that can be mobile. In both cases, the model is today not sufficiently offense-conservative—tending to offset the above two biases.

The reader will have perceived by now that the criticisms of the APM have been chiefly directed at its use in the JSBS, which found equal cost forces that included the B-1 to be significantly more effective than those without the B-1. An additional criticism of that application was that alternative mixed forces were evaluated, so that in effect nothing (except estimated ten-year force costs) was held constant. However, mixed forces are realistic.

In retrospect, one may argue that the JSBS could have had greater face validity and therefore salability if it had also tested pure forces. It seems clear that such an approach, though less realistic, would have shown: (1) even greater differences between the respective bombers, since APM runs have shown that B-52s perform better (in targets killed) in combination with B-1s than in a pure force; and (2) that forces combining cruise missiles, SRAMs and gravity bombs perform better than forces with ALCMs only.

CONCLUSIONS

The APM cannot be used to predict the outcome of the bomber component of a strategic nuclear war. Once again, the absolute output of a model is of dubious value.

The APM can be used to compare alternatives, including:

—Alternative forces,

—Alternative new weapons or weapon characteristics,

—Alternative tactics, including limited attacks,

The APM can be a powerful learning tool for those studying the above kinds of choices.

The APM Mission Planner cannot in its present form be used for actual mission planning, since it does not assure optimum plans. It may, however, aid SAC in studying alternative plans.

The APM has had long and extensive use for a wide variety of studies. It must therefore be assumed to have generally broad acceptance in the Air Force. It is true that its use in the JSBS generated widespread controversy and that not all of the major findings of the JSBS were followed in decision-making, but it is not clear that the use of any other model or analytical study would have led to a different outcome.

As noted above, the APM has been provided to SAC, where it will continue to be both maintained for widespread use and improved as defects can be remedied. Efforts are continuing to assure compatability with the Advanced Missile Model (AMM), to permit its use in larger strategic nuclear war studies.

Efforts are under way, by contract with General Research Corporation, to develop means of approximating APM findings by means of analytical models. It should be emphasized that, to the extent that these efforts may succeed, they will not demonstrate that simpler analytical tools could have achieved the same results at great savings in resources expended. Analytical solutions that reflect the learning form, and in particular coefficient values derived from, an accepted simulation can be used with far greater confidence than those derived purely deductively.†

Questions

1. How could APM findings be related to a ballistic missile attack model?
2. How would the answer to Q. 1 change if there were ABM deployments?
3. Do you think the APM adequately models ECM-ECCM effects?
4. Does the APM adequately account for differences in the penetrativity of different bombers? Does it adequately reflect the effect of different munitions on the effectiveness of bombers?
5. If you were assigned to the APM project—and could choose your task (!)—what aspect of the model would you choose to work on? Why?

References

1. "Measures and Trends: U.S. and USSR Strategic Force Effectiveness," Draft Interim Report of Santa Fe Corporation for the Defense Nuclear Agency, February 1978.
2. *Congressional Record,* p. S6618, May 6, 1976.
3. "Advance Penetration Model (APM)," Briefing Headquarters, USAF Assistant Chief of Staff for Studies and Analysis, 1978.
4. Archie L. Wood, "Modernizing the Strategic Bomber Force without Really Trying—A Case Against the B-1 Bomber," *International Security,* Vol. 1, No. 2, Fall 1976, pp. 98–116.

†An interesting example of the fallacy of assuming otherwise is to be found in a footnote[4] to an article by Archie L. Wood, one of the critics of the JSBS and the B-1, in which he said:

> Proponents of the B-1 were especially critical of the simple "subtractive" penetration model used in *Modernizing the Strategic Bomber Force.* At least one DoD spokesman prefers an "exponential" model, of the form $p = e^{-x}$, which he said was validated by more complex calculations done in the DoD Joint Strategic Bomber Study. In this equation, p is the penetration probability and x is the ratio of potential air defense intercepts to penetrators. The series of expansion of e^{-x} is
>
> $$1 - x + \frac{x^2}{2!} - \frac{x^3}{3!} + \ldots$$
>
> The "subtractive" model consists of the first two terms of this expansion. For small values of x (the only condition in which the subtractive model was applied) the third and subsequent terms are small compared to the first two terms and to other uncertainties in such calculations, and can be ignored.

While his mathematical point is correct, what Wood's comment obscured was that his subtractive model was a set of closed form equations that fixed the number of air defense intercepts for all cases, independent of number and type of penetrators, thus assuming infinite defense resources and ignoring penetrator characteristics and penetration aids.

CHAPTER V

The Theater Tactical Air Campaign

THE PROBLEM

Chapter I touched on the complexities of tactical air warfare, especially as it moved from the simplicity of early World War I "Red Baron"-type one-on-one dogfights through few-on-one and few-on-few to many-on-many. This chapter will elaborate on the current status of attempts to understand and model the problems of tactical (conventional) air campaigns. The air campaign involves many missions that go far beyond air combat to include those of aircraft in ground support and attack and of ground weapons opposing them—close air support (CAS), battlefield defense, barrier air patrol, offensive counter-air and interdiction, defensive counter-air (including surface-to-air missiles, or SAMs), electronic warfare (ECM, ECCM), and so on.

The case studied is that of the Air Force Studies and Analyses TAC WARRIOR model designed to aid in making weapons choices and force structure decisions, as well as force employment concepts for given force structures. But it must not be lost sight of that, while aircraft must attack each other and defend themselves, their primary reason for being (in theater warfare) is to support the ground forces, and the ultimate figure of merit is ground attrition, especially of tanks and other vehicles killed in CAS and battlefield interdiction. This is, incidentally, institutionally difficult—pilots become aces for shooting down planes, not shooting up tanks. Nevertheless, aircraft do not take and hold territory. As long as deterrence of nuclear war is an overriding national objective (see Chapter VII) and the Air Force must be prepared for conventional war, CAS will remain a primary mission, albeit difficult to understand and to model, as will become doubly evident in the review of the theater nuclear war modeling problem in Chapter VI.

If the basic objective of tac air is to support the ground forces, the above comments suggest the extensive complex of roles played by tac air in moving toward this objective. Clearly, the struggle for air superiority is a dominant theme. Douhet perceived almost sixty years ago that if "command of the air" could be achieved, the Air Forces would then be free to attack ground targets unimpeded. This involves trade-offs of a complexity that he did not perceive. His scenario was one in which one side understood his thesis and prepared to achieve command of the air, while the other side did not.[1] We are looking here, however, at a symmetrical scenario. Both sides perceive the problem and prepare to meet it in similar ways.

In the first place, ground support may mean CAS at the FEBA or combat area (since the FEBA is unlikely in the future to be of the classic linear and well-defined form) as well as to the rear. How far in the rear is a good question. For a few tens of kilometers back, the flow of men and material toward the front provides a potential class of targets that will affect the battle in hours or days. Farther back, staging areas, ammo dumps, vehicle parks (dispersed?), repair depots, bridges and other transportation nodes, etc., are targets the destruction of which can affect the battle in days to weeks. Deeper interdiction can attempt to prevent longer-term reinforcement and indeed to destroy the base of both support and morale on which any army ultimately depends.

Thus, there is continuum back to the destruction of the industrial base in the homelands of the combatants and/or of their suppliers. For evident reasons, we will omit this last stage here:

1) history suggests that bombing, per se, helps rather than destroys morale, so long as the armies are not destroyed or clearly on the way to defeat;

2) massive industrial targeting, only moderately effective with HE, is likely to take place in a future war only with nuclear weapons, a topic arbitrarily ruled out here and taken up in Chapters VII and VIII.

The menu of ground targets is, then, rich. But the choices cannot be made solely in terms of predicted impact on the ground battle—a difficult and uncertain enough task in itself. At each stage, the enemy will attempt to defend themselves—all the while trying to do the same harsh things to our side.

Given that resources are scarce—and virtually fixed from the point of view of the commander of a battle, every use of an aircraft is at the expense of some alternative use. CAS means less interdiction to the rear. Both mean less defense vs. enemy CAS and interdiction, less suppression of his defenses (destruction of ground defenses, C^3, and air bases), and less resources with which to defeat his aircraft in combat.

This is an opportunity-cost view of the trade-offs. These trade-offs can also be looked at in terms of their synergistic effects. The more successful the defense suppression and achievement of air superiority, the easier and more successful will be the CAS and interdiction missions, and the less successful will be the enemy in these missions.

Every planner and commander assumes that the answer to the choices outlined above is a mix—of missions and of equipment. This Chapter will describe one ambitious attempt, not to optimize this mix but to aid planners in comprehending the problems and in testing alternative force and employment mixes.[2]

AIR-TO-AIR COMBAT SIMULATION METHODOLOGY

Before discussing TAC WARRIOR in some detail, a few words may be in order about the types of air-to-air combat simulations that have been or are in use. These differ in level of detail and scope, although less detail does not always imply wider scope. The following examples are presented in roughly ascending order of scope. Some of the narrowest will be recognized later as being used to compute inputs to the very-broad as well as detailed TAC WARRIOR.

The narrowest in scope as well as most detailed air-to-air combat simulations are the so-called "fly-out" and "endgame" simulations. Fly-out models simulate in great detail the flight of an air-to-air missile in one-on-one combat situations. The positions of the launching and target aircraft are specified at time of launch. The missile is fired against the target, which is allowed to execute preprogrammed maneuvers. The fly-out simulation then accounts for the performance of the propulsion and guidance systems of the missile. Minute details of the missile electronics and aerodynamics can be simulated. The models may typically consume five to ten seconds of computer time for every second of missile flight. The output is a list of all the parameter values describing the state of the missile and when it made its closest approach to the target. Every missile in the USAF and Navy inventories has a fly-out model associated with it.

Given the above description of the state of the missile at its point of closest approach, an endgame model can determine the probability of target kill (P_k). An endgame model contains detailed descriptions of the target as well as the missile's fuse/warhead subsystems. Warhead detonation and target damage are calculated to estimate P_k. Each evaluation can consume ten seconds or more of computer time. SCAN, SHAZAM, ENDGAME, and OASIS are all examples of endgame models.[3]

Somewhat broader in scope are the "Energy/Maneuverability" models. These eliminate the missile performance detail by assuming values of P_k as a function of firing range and angle as well as launcher velocity and target maneuver, based on the findings of the above models. Adversary aircraft are modeled in three dimensions and have six degrees of freedom of motion. The aircraft maneuver to maintain the maximum specific energy, on the assumption that the aircraft with the greatest specific energy has the advantage. This is a "decision rule" approach under which the simulated pilots will always make the same decision in the same specific energy situation, regardless of anything else that is happening. More complicated decision trees can be included. PAC AM, TACTICS II, and TAC AVENGER (used for TAC WARRIOR inputs) are examples.[3]

The scope can be broadened by including the "man-in-the-loop." These simulations are similar to the above, with the maneuver decisions made by a human "pilot" interacting with the model. These simulations may actually involve a pilot being placed inside a projection dome on which the computer projects images of the background and the target on the dome in real time. The NASA "Differential Maneuvering Simulator" (DMS) is an example.[4]

The further broadening from one-on-one to many-on-many is being developed under the "value-driven decision theory" approach. The effects of information used by the combatants to make decisions is accounted for in the simulation of their interaction. Each pilot has his own set of values. These values may change, depending on the level of decision—e.g., wing man vs. flight leader. The values may also change according to the overall combat situation or on the basis of orders received by radio. The result is that at any instant of time each pilot has a "mental model". This model determines the value resulting from each possible course of action, based on the projection into the future of the results of that action. The course of action with the highest value is selected (see Figure 24). TAC BRAWLER is currently developing this approach (largely as a result of TAC WARRIOR experience), although elements of the approach are used in the NASA DMS simulation.

Also of broad scope is the "maneuver conversion" model. The engagement is modeled as a collection of states. Individual aircraft and their activ-

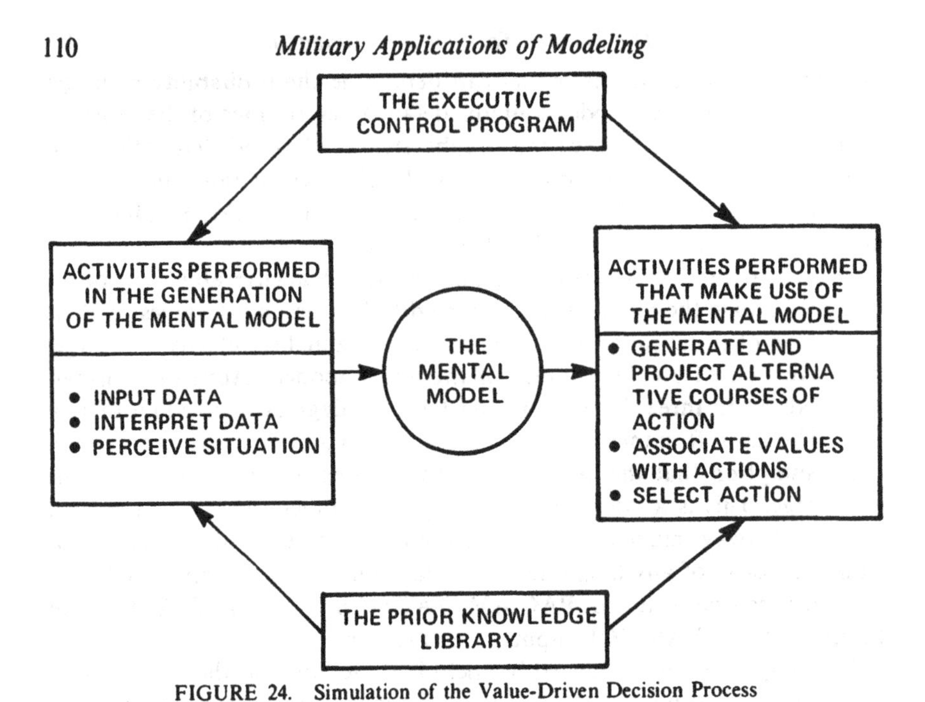

FIGURE 24. Simulation of the Value-Driven Decision Process

ities are not explicitly described. Rather, the situation is viewed as a semi-Markov process in which there is a certain probability of transitioning from one state to another. These probabilities are determined from other, more detailed models or from test range data. This is the model for the air-to-air combat portion of TAC WARRIOR, described below.

Broad scope and considerable realism and achieved in the Air Combat Maneuvering Instrumentation/Range (ACMI/R) at Nellis AFB, Nevada. Here, actual aircraft are flown by human pilots who are free to maneuver as they see fit, subject to safety regulations. Missiles are simulated via a real-time data link between all combatant aircraft and a ground-based computer. When a pilot "fires" a missile, the computer utilizes a low-fidelity fast-fly out simulation to fly it to the target aircraft. Given the terminal engagement conditions, tables are entered to determine whether a kill is achieved. If so, the "dead" pilot is immediately instructed to leave the combat arena.

It has been suggested[2] that a major determinant of aerial combat outcomes is the experience and proficiency of the pilot. This can be taken into account the ACMI/R simulation. This is, of course, a very expensive simulation, but presumably the cost is chargeable to training.

There is still one basic problem: the ACMI/R physical simulation is essentially risk-free. In these circumstances, there may be strong incentives for the pilot to become an ace, or at least less disincentive than in war. This suggests that the observation that experience is a major determinant of outcomes should add caution, or instinct for self-preservation, to proficiency levels in describing the value of experience. It was widely observed in the two World Wars that pilots who came back from their first few missions were likely to survive the full fifty missions which were the nominal maximum (in large part because of the observation or belief that combat fatigue can eventually offset the value of experience). This seems to be the Air Force version of the old saying, "Old soldiers never die." In any event, we are forced back to the conclusion of Chapter I, that the only truly valid simulations are actual wars, preferably small. Lacking this analytical tool, and retaining our desire to rely on computer simulations, let us proceed to an examination of the TAC WARRIOR air campaign.

TAC WARRIOR

TAC WARRIOR is a large, complex computer simulation developed in 1974–75.† The simulation describes mathematically and logically the encounter between two tactical air forces in a theater of operations. It proceeds deterministically for ten days through fixed time increments of about one minute, i.e., about 14,000 time steps. Figure 25[5] shows the missions simulated in TAC WARRIOR. They are perfectly symmetrical, in model capability, between Blue and Red, although of course the data bases will reflect asymmetries in numbers and types of equipment, airbases, and employment doctrine. The chart should be read from the bottom up. At bottom center is the ground battle at the (increasingly) vaguely defined FEBA, or forward edge of the battle area. Each side has the opportunity to provide close air support (CAS) from its air bases, which means essentially attempting to kill enemy tanks and other armored fighting vehicles (AFVs). As we will see when we come to describe the methodology, the model keeps count of the probable kills by each side in terms of numbers of notional vehicles, having no way to distinguish vehicle types, since it does not simulate the ground battle. Each side can also send up fighter aircraft to try to intercept the CAS planes, and each side can maintain a barrier combat air patrol (BARCAP) to engage the battlefield defense before it can interfere

†By the Assistant Chief of Staff, Studies and Analyses (AF/SA), Fighter Division, Headquarters United States Air Force, with documentation prepared by Control Analysis Corp.[5,6]

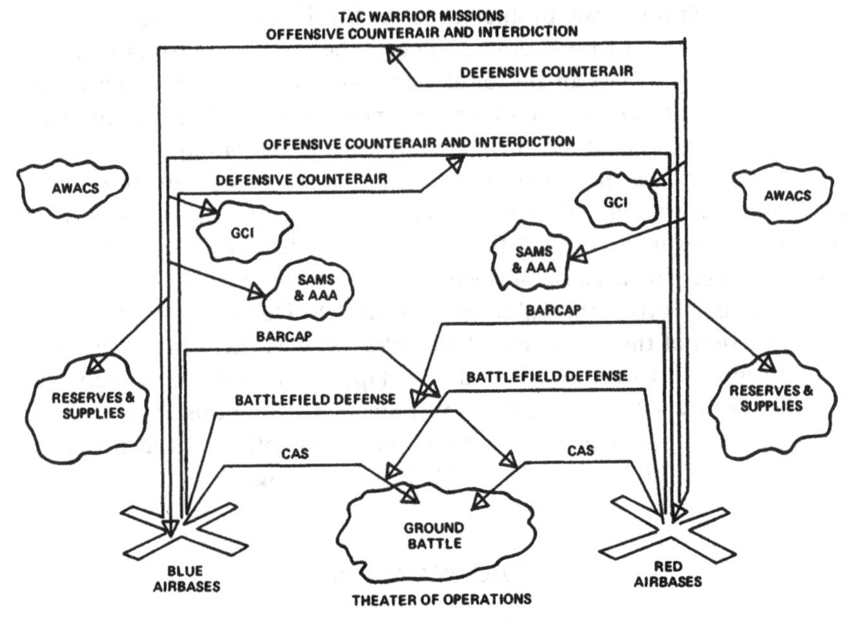

FIGURE 25. Theater of Operations

with the CAS aircraft. These three first-line missions, CAS, battlefield defense, and BARCAP, are collectively referred to as the FEBA missions.

The model has the capability for offensive counter-air and interdiction missions by Red or Blue fighter-bombers to attack enemy air bases, as well as reserves and supply depots. These missions, along with the defensive counter-air that attempts to stop them, are called the "Hammer" missions. Note that the offensive counter-air and interdiction aircraft could be targeted against the ground control intercept (GCI) radars, surface-to-air missiles (SAMs), and anti-aircraft artillery (AAA), in order to reduce effectiveness against subsequent attacks on airfields or reserves.

In addition to GCI radars, interceptor radars, and visual search, to enable opposing aircraft to find and attack (or evade) each other, the model has the capability for either side to employ airborne warning and control systems (AWACS) for acquisition and for vectoring of interceptors. However, there are no arrows pointing to the AWACS in Figure 25. No one has as yet successfully modeled the attack and defense of AWACS for TAC WARRIOR. The AWACS aircraft (a modified Boeing 707 for Blue) may be quite vulnerable, but they also patrol well behind the battle lines, have very-long-range radars (covering great areas because they operate at medium-high altitudes), and can control their self-defense (to which interceptor aircraft may be dedicated).† Two or three patroling, or orbiting, AWACS aircraft can be assumed on each side throughout the campaign. This means rotation with reserve aircraft available on the ground. One way to protect the AWACS, therefore, would be the maintenance of additional aircraft ready to replace destroyed AWACS aircraft. If they are considered invulnerable in the air, they could always be attacked on the ground, though they may be very deep-based and such attack is not modeled in TAC WARRIOR.

INPUT MODELS

TAC WARRIOR is an hierarchical model in the sense that it depends for many of its inputs on a set of lower-level models. The latter are a system of high-resolution "micro" models, as shown in Figure 26. These are detailed engineering, one-on-one models of individual components of TAC AIR, designed to evaluate the effectiveness of tactical air weapon systems within specific roles and missions, as well as models describing aircraft turn around, maintenance, detection, etc. The term "micro" refers to the level of analysis, or suboptimization, of the sub-models, not to their size or complexity. As we will see, they may be very large, complex models in themselves. But they are not necessarily run anew with each TAC WARRIOR run.

Air-to-Air Effectiveness

Starting at top left in Figure 26, TAC AVENGER (for *T*actical *A*ir *C*apabilities, *AV*ionics, *EN*er*G*y maneuverability *E*valuation and *R*esearch) is a digital computer simulation of two aircraft in both close-in (CI) and beyond-visual-range (BVR) air duels. In this simulation, each aircraft maneuvers in three dimensions; each pilot reacts on a second-by-second basis to the maneuvers of the opponent; and each pilot expends ordnance against the other aircraft as opportunities occur. The aircraft and weapons per-

---

†Note that an AWACS aircraft can not only carry a much larger (higher power) radar than an interceptor aircraft attacking it, but, with a look-down capability, it can scan to a theoretical horizon of about 245 nautical miles from a 30,000-foot altitude:

$$D = (2h)^{1/2},$$

where D = horizon distance in nautical miles and h = antenna height in feet above the earth.[6] Boeing claims only 212 nautical miles, but points out that it can see aircraft significantly beyond the horizon unless they are flying at "low altitude."[7]

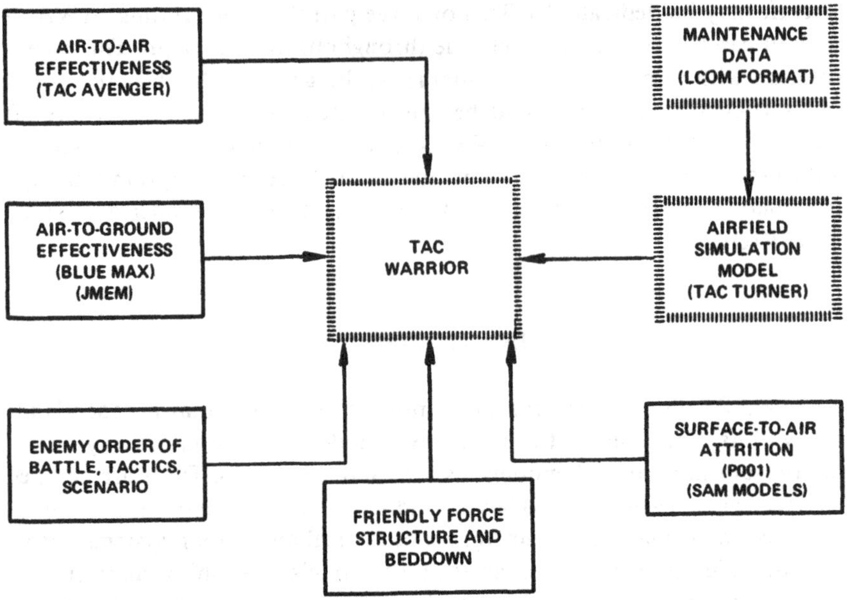

FIGURE 26. Model Inputs

formance are described in detailed engineering data. Individual aircraft tactics are selected from a range of reasonable choices, based upon the tactical situation, the relative performance capability of the aircraft, and pilot preferences. These individual pilot preferences were derived from empirical, real-world data with respect to the avionics, energy maneuverability, and weapons-to-fighter effectiveness of given aircraft.

TAC AVENGER development was initiated in 1967 by Air Force Studies and Analyses, on the basis of earlier work on the North American FANTAC model. It uses a FORTRAN IV language program with a GE-635 computer system. It utilizes 44K words of core memory, with 94 routines having 19,000 source statements. The average CPU running time is 2.5 minutes for each 5 minutes of simulation.

Air-to-Ground Effectiveness

The next supporting model is BLUE MAX (Variable Airspeed Flight Path Generator), a model which generates a variable speed flight path. Each aircraft is entered with its weight, wing area, and lift and drag coefficients stated as a function of Mach number. Thrust is entered as a function of Mach number and altitude. Five types of maneuvers are constructed: navigation; base; roll-in/dive; pull-off; and recovery/jink legs. The methodology is based on the assumption that the pilot is able to exercise control of only three variables: pitch, roll, and power. The rate of change in magnitude of these variables is a function of the type of maneuvering encountered. The model treats the aircraft as a point mass, solves for the lift, drag and thrust to determine the three orthogonal accelerations in the body axis, and then transforms these into an inertial reference. The inertial acclerations are then used to update the velocity and position of the aircraft.

The Blue Max flight paths are used in conjunction with *Joint Munitions Effectness Manuals* (JMEM)† to determine the effectiveness of air-to-ground attacks and which ground targets are to be treated as destroyed. The flight paths are also inputs to the AAA and SAM attrition models.

BLUE MAX was written in FORTRAN for the GE-635, with 24K words of core memory, but is now stored on MULTICS. The average time for building a single flight path is about 30 minutes when using an interactive operation.

Red and Blue Force and Supply Data and Scenarios

The lower left and bottom center boxes refer to direct imputs to TAC WARRIOR, giving Blue and Red numbers and characteristics in the force structure, and specified addition to specifying the characteristics and "states" (below) of four types of friendly and enemy penetrator and defender aircraft and three types of air-to-ground attacking aircraft on each side, a tremendous amount of information is provided on the Blue and Red airbases. Since TAC WARRIOR must assess damage to these airbases tactics and scenarios. These involve a tremendous amount of detail, as will be further illuminated in the discussion of the TAC WARRIOR simulation itself. For a capability to generate sorties from them as the air campaign progresses, great detail is required. No aircraft should be assumed to take-off from a runway turned to rubble, without fuel, etc. (as could occur in earlier models, in which sortie rates were independently supplied). Accordingly, for each airbase, there are inputted five types of munitions, aircraft shelters, maintenance shelters (larger, for the performance of major [level II] maintenance), POL, and runways.

†These are manuals published by the Joint Technical Coordination Group for Munitions Effectiveness (61 JTCG/ME-4, 15 September, 1975) and reflect interservice (Army Materiel, Naval Materiel, Air Force Logistics and Air Force Systems Commands) agreement on the battlefield effectiveness of munitions, taking account of target vulnerability, weapon characteristics (damage mechanisms and explosive forces), and delivery accuracy.

Supplies, Maintenance, and Sorties

In addition to the states of each of these components for an aircraft returning to base or taking-off again, a great deal of information is required on maintenance facilities, and repair frequencies and times. This brings us to the upper right hand box, which represents the LCOM (Logistics Composite Model) data base. In this data base, component repair networks represent the repair sequence associated with one component of an aircraft. The component fails as a function of sorties flown or mean sorties between failure (MSBF). Repair includes mean task-time, type of distribution, variance, and maintenance crew size and specialty, as well as failure rates for selected spare parts, repair of parts in the maintenance shops, and reorder from the depot if the part cannot be repaired on base.

LCOM is a Monte Carlo simulation that models the work centers that contribute directly to sortie generation. It accounts for the impact of resource quantities on the ability of an organization (airbase) to generate sorty-ready aircraft—hence its appropriateness to measure the effect of base attrition on the air campaign in TAC WARRIOR.[8,9]

Typically, two to three hundred individual aircraft components may be included for a single aircraft weapons system. Not surprisingly, therefore, the LCOM model is huge and cumbersome. Approximately 100-110K core words and one and one-half to two hours CPU time are required for one LCOM run. Moreover, the same LCOM was established as a Monte Carlo simulation, one run represents just one data point. TAC TURNER, in the next box, was therefore developed to simplify the LCOM data base by replacing each detailed network with its expected value and variance. TAC TURNER is an event simulation model of aircraft turnaround activities on a tactical airbase. The model determines surge sortie generation capabilities for various tactical aircraft when constrained by airbase resources (e.g., maintenance, manpower, spare parts, POL, munitions, and aircrews). Turnaround functions include arming/dearming, battle damage repair, unscheduled maintenance repair, cannibalization, attrition, refueling, weapons loading, and WRM (war reserve material) resupply.

TAC TURNER was developed by Air Force Studies and Analyses in 1975. Written initially in GESIM (GE-635 General Simulation Compiler), it was converted to the more flexible General Purpose Simulation System (GPSS) language for the IMB 370 computer. A typical run on the IBM 370 requires 256K bytes of core memory and two minutes of CPU time. Comparative runs of TAC TURNER with LCOM indicated a 95 percent saving in core hours (core capacity requirements times computer time), with a five percent variation in resulting average sortie generation rates.

Surface-to-Air Attrition

We come finally to the lower right-hand corner box for surface-to-air attrition. It will be recalled that Blue Max determined the flight path of the various types of aircraft attacking various types of ground targets (see Figure 25). The P001 Model computes the single-shot probability of kill (SSPK) of a target aircraft flying through the coverage of antiaircraft artillery (AAA)—presently, the Soviet ZSU-23-4, a radar-controlled "quad," or four-barrelled, 23mm cannon. Aircraft attrition along an entire flight path through AAA envelopes (space of coverage) is obtained through accumulation of the SSPKs. The major portion of the program is concerned with the analysis of the various sources of random error that influence the effectiveness of the AAA. After assessment of these errors, the vulnerable area of the aircraft is located within the total distribution of AAA trajectories, and the total probability of kill is computed. P001 was developed by the Air Force Armament Laboratory, and the software was documented by Armament Systems, Inc. P001 is used on the MULTICS computer. There are approximately 1,400 FORTRAN source statements, and the running time for some typical encounters is approximately 5 minutes.

Surface-to-air missile (SAM) attrition is calculated in the family of TAC ZINGER models, of which 2, 3, 4, 5, 6, 8, and 7/9, are at time of writing in use. The different versions of the model apply to different types of SAM systems, (the Soviet SA-2, etc.), that is, those using radar, optical, or infrared (IR) sensors, and command guidance, active, semiactive, or passive homing, etc. TAC ZINGER is a one-on-one computer simulation of guided air defense missile systems. It is a Monte Carlo model of the interaction of a maneuvering aircraft flight path with a guided missile path. The interaction takes account of target signatures, atmospheric attenuation, seeker sensitivity, guidance dynamics, missile kinematics capabilities, target-vulnerable areas, aircraft ECM (electronic countermeasures), and missile warhead fusing and lethal radius.

TAC ZINGER started with a three-degrees-of-freedom model, an improved version of the U.S. Army Missile Command Model, "3-dimensional Air Defense Kinematic Launch and Intercept Boundary Computer Program." TAC ZINGER 7/9 added a five-degrees-of-freedom model developed in Air Force Studies and Analysis. All versions are written in FORTRAN and require some 30,000–35,000 words of core storage. Using the MULTICS computer, one-on-one simulations may require only thirty seconds of computer time, but the most complex scenarios may run from ten minutes to several hours.

One final external input is called TAC GROPER. This is not shown on

Figure 26, because it is not really a computer model but simply a set of exponential equations used to calculate detection probabilities wherever enemy aircraft are approaching each other, using parameters for the detection sector and range for a given aircraft, the number of such aircraft in a group, and the radar and visual signatures of the enemy aircraft in the particular pairing.

THE METHODOLOGY

The concept of the TAC WARRIOR simulation methodology is the resource "loop." A loop refers to a closed module, or system that contains all the states, or levels, in which a particular resource can exist throughout the course of the campaign. Figure 27 gives a very simplified example of a loop that represents an aircraft performing a mission. Seven states are illustrated, although actual loops may contain up to 100 states. Each state represents a particular disposition of elements of the resource. At any given time, each state will contain a certain number of resource units (remembering that Δt is about one minute). There may also be off-line dummy states that amount to counters for simulating such things as numbers of take-offs or sorties at various stages (note box for numbers killed at right of Figure 27).

Flow Rates

A pair of states may be joined by a directional "flow path." These flow paths indicate the possible state-to-state transitions that a resource element

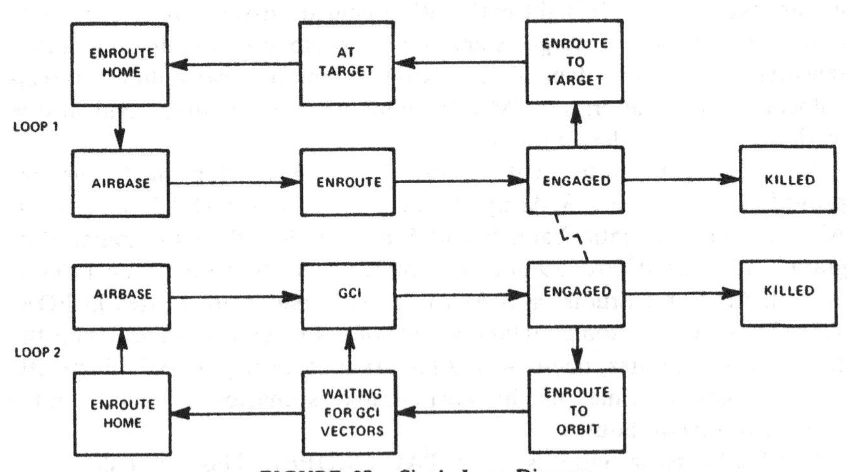

FIGURE 27. Single-Loop Diagram

can make. The actual transitions are controlled by flow rates across the flow paths. These flow rates are defined by flow-rate differential equations, derived from the input models. The exact forms of the equations depend on the physical situations being modeled. For example, the transition from "en route home" to "airbase" could be controlled by an equation that is a function of only the number of elements en route home and a mean flight time assigned to the return leg. The differential equations for the transition between any of these kinds of states have the general form (referred to as the "X-dot" equations):

$$\dot{X}(t) = N(t)/T,$$

where

$\dot{X}(t)$ is the instantaneous flow rate of resources out of a state at time t,
$N(t)$ is the level of resources in the state at time t, and
T is the mean time a resource element spends in the state.

This is the simplest form of the basic differential equation of the model. For the transition from state i to state j, it would be written

$$\dot{X}_{ij}(t) = N_i(t)/T_i,$$

stating that the flow rate is proportional to the current resource level in state i. This represents the assumption that the resources in the given state are evenly distributed with respect to time.

A higher level of complexity occurs in many X equations when T is in itself a complex function. For example, the unscheduled maintenance flow rate on an airbase (see Figure 28) would be

$$\dot{X} = N(t)/MT,$$

where

$\dot{X}$ is the instantaneous flow rate of aircraft out of unscheduled maintenance,
$N(t)$ is the number of aircraft in unscheduled maintenance at time t, and
MT is the average time to complete unscheduled maintenance defined as

$$MT = K_1 R^{C_1} T^{C_2} S^{C_3},$$

in which

K_1, C_1, C_2, and C_3 are constants determined by regression on TAC TURNER results,
R is the ratio of spare parts and maintenance personnel remaining to aircraft remaining,
T is the time elapsed since start of flying aircraft, and
S is the sorties per day per Primary Assigned Aircraft (PAA).

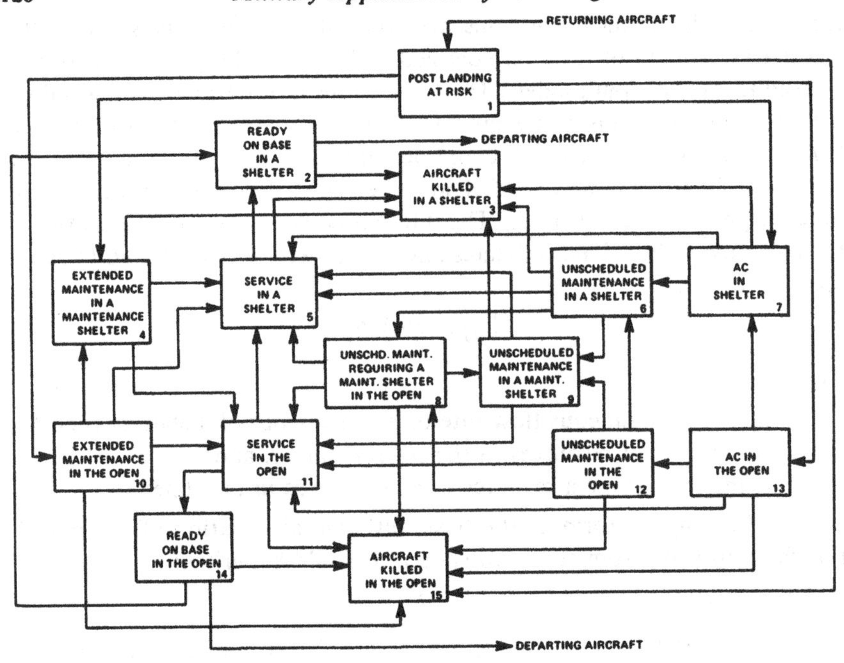

FIGURE 28. Air Base Flow Diagram

Two loops (one Blue and one Red) may be connected by an engagement flow rate, or attrition rate—the dashed line in Figure 29. Attrition equations can become very complex. As a sample, for a beyond-visual-range (BVR) duel the rate at which Red aircraft are killed by Blue aircraft, may be represented as:

$$\dot{X}_{ij} = R_1\left[\sum_{i=0}^{k} \mathrm{RFK}_i\binom{B}{i}(1/R)^i(1 - 1/R)^{B-i} + \mathrm{FRK}_{k+1}\, E_{k+1}\right]/E[T],$$

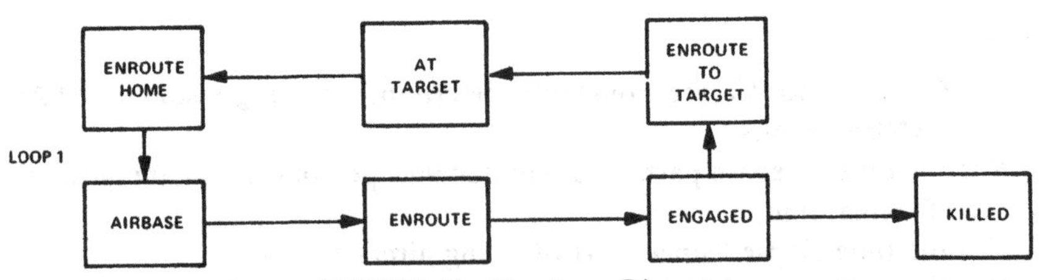

FIGURE 29. Two-Loop Diagram

where

FRK is the fractional rate of kill,
R is the number of Red aircraft killed per Blue aircraft engaged per minute,
B is the number of Blue aircraft engaged, and
E is a correction term to account for probability mass not accounted for in states $i = 0$ and $i = k$.

Rates of Change-of-State

Thus far, we have been discussing the flow rates between states, represented by $\dot{X}$ equations. The equations for the rates of change in these states are called $\dot{N}$ equations, of the form

$$\dot{N}_i(t) = \sum_k \dot{X}_{ki} - \sum_j \dot{X}_{ij}$$

where

$\dot{N}_i$ is the rate at which state i is changing at time (t),
$\dot{X}_{ki}(t)$ are the flow rates into state i at time t, and
$\dot{X}_{ij}(t)$ are the flow rates out from state i at time t.

These N equations are solved by Euler's method:

$$N_i(t + \Delta t) = N_i(t) + \dot{N}_i(t)\Delta t.$$

Going back to Figure 28, for example, the change in state 9 for a given Δt would be the sum of the rates of flow from states 6, 8, and 12, minus the rates of flow to states 3, 5, and 11—in short, plus all $\dot{X}_{ki}$ and minus all $\dot{X}_{ij}$ where $i = 9$, as follows:

Plus	*Minus*
$\dot{X}_{6\ 9}$	$\dot{X}_{9\ 3}$
$\dot{X}_{8\ 9}$	$\dot{X}_{9\ 5}$
$\dot{X}_{12\ 9}$	$\dot{X}_{9\ 11}$

Complexity

Some idea of the size and complexity of the model may be gained from the fact that there are some 17 Red and 17 Blue aircraft loops (a total of 9 for FEBA missions, 8 for Hammer missions), plus loops for GCI and AWACS on each side. The loops may have from as few as three states to as many as 104 (Figure 30). The loops listed have a total of 2,564 aircraft resource

- AIR BATTLE OVER ENEMY TERRITORY
  - 4 SEPARATE TYPES OF FRIENDLY PENETRATOR AIRCRAFT WITH 82 STATES EACH
  - 4 SEPARATE TYPES OF ENEMY DEFENDER AIRCRAFT WITH 104 STATES EACH
  - ENEMY GCI/AWACS PAIRINGS WITH 3 STATES
- AIR BATTLE OVER FRIENDLY TERRITORY
  - 4 SEPARATE TYPES OF ENEMY PENETRATOR AIRCRAFT WITH 82 STATES EACH
  - 4 SEPARATE TYPES OF FRIENDLY DEFENDER AIRCRAFT WITH 104 STATES EACH
  - FRIENDLY GCI/AWACS PAIRINGS WITH 3 STATES
- FRIENDLY CLOSE AIR SUPPORT IN VICINITY OF MAJOR GROUND BATTLE
  - 9 SEPARATE TYPES OF FRIENDLY AIR-TO-GROUND AIRCRAFT WITH 92 STATES EACH
  - 3 SEPARATE TYPES OF FRIENDLY AIR-TO-AIR AIRCRAFT WITH 58 STATES EACH
  - 3 SEPARATE TYPES OF ENEMY AIR-TO-AIR AIRCRAFT WITH 64 STATES EACH
- ENEMY CLOSE AIR SUPPORT IN VICINITY OF MAJOR GROUND BATTLE
  - 3 SEPARATE TYPES OF ENEMY AIR-TO-GROUND AIRCRAFT WITH 92 STATES EACH
  - 3 SEPARATE TYPES OF ENEMY AIR-TO-AIR AIRCRAFT WITH 58 STATES EACH
  - 3 SEPARATE TYPES OF ENEMY AIR-TO-AIR AIRCRAFT WITH 64 STATES EACH

FIGURE 30. Level of Detail

states. Over 5,300 $\dot{X}$ differential equations are required to describe the interactions between states.

In addition, each aircraft loop has associated with it an airbase portion of the loop (Figure 28, above). Each airbase has associated with it loops representing the following resources:

—2 types of munitions, with 3 states each;
—aircraft shelters, with 3 states;
—maintenance shelters, with 3 states;
—POL with 3 states; and
—runways, with 4 states.

These resources add 760 states, for a grand total of 3,324, each requiring an $\dot{N}$ equation.

As noted earlier, the simulation proceeds in one-minute time intervals, or steps, over a period of ten days. The selection of Δt equal to one minute was made early-on on the basis of a series of sensitivity tests in which Δt started as small as 10 milliseconds. One minute was found to represent the knee of the curve beyond which increases in the time increment led to rapidly increasing instability of the solutions. Differential equations may have to be recalculated for each Δt.

Some reasonable simplifications have made possible reductions in the implied 14,400 recalculations of each of the states of the loops.

Simplifications

One of these simplifications arose from the recognition that not all activities are continuous throughout the ten-day campaign. Aircraft cannot fly continuously and there are therefore considerable periods of inactivity in given loops. For example, Red air raids in the model might start at x00 hours. If these raids generate a response by Blue interceptors, then at the end of the raid and defense the Blue aircraft return to their base. TAC WARRIOR then puts the planes through the "turn-around" process on their base, using TAC TURNER-generated factors. The planes can only be in one of the states on base at a time, from "post-landing at risk" through several maintenance, turn-around and in-service states, before they are "ready-on-base" for possible take-off again. The steps in this low-activity period are treated as stochastic, or probablistic, in a Markovian process, or "chain," in which the conditional probability of the next step in the process is dependent only on the preceding state, i.e., independent of any past states that led up to this state.[10] It was shown during the development of TAC WARRIOR that for periods of low activity (nighttime, between missions) the states could be adequately represented by a Markovian chain with stationary transition probabilities and known initial states, thus permitting a closed-form matrix solution of the states involved, with significant computational savings.

Further simplification is achieved by limiting the air duels between the several categories of missions and types of aircraft (see again Figure 25) to four or five stages, adding additional stages, up to eight, from off-line presimulation sub-routines called STOCAS and STOCA1. STOCAS and STOCA1 make stochastic precalculations of the expected kill-rates and engagement times for multi-stage duels.[10] The basic data for these calculations are engagement times and probabilities of kill derived from TAC AVENGER for one-stage one-on-one duels between all pairs of aircraft types under the conditions being modeled.

Two basic types of duels are included: close-in (CI) and stand-off or beyond-visual-range (BVR) duels.

The CI duels are generally modeled as two-stage duels that start with match-ups of one-on-one, two-on-one, three-on-one, or four-on-one. An exception occurs when one pilot views himself as being hopelessly outnumbered and attempts to avoid the engagement, in which case he may be subject to a BVR attack as he flees. Whether he escapes will depend on the timing of his break-off, his maneuverability, and his ECM capabilities vis-a-vis those of the opposing aircraft/missile type.

In BVR duels, it is realistic to postulate longer periods of contact and therefore more potential stages of the duel. In this case, therefore, four stages are modeled, each starting with match-ups of one-on-one, two-on-one, three-on-one, or four-on-one. The user may elect to skip one or more of these stages.

The reader may well ask, why confine modeling capability to one-, two-, three-, or four-on-one, since duels of few-on-few and many-on-many may be readily envisioned? The answer at this stage has been that the larger group of aircraft in an engagement can generally enforce one of these kinds of duels in any one stage of a multi-staged duel. Two factors suggest that the side which is numerically superior and, indeed, the side which can really enforce a given type of duel, may change, in both the CI and BVR cases: first, the exchanges are stochastic, and, in individual engagements, the results will vary (for the expected values used for the TAC WARRIOR simulation, repeated Monte Carlo runs are made in the preprocessors); secondly, superiority of one or more aircraft and/or air-to-air missiles on one side, expressed in higher kill probabilities (P_ks), may eliminate superiority in numbers on the other side.

It is interesting to note that, apart from deviations from expected values in individual cases, the attrition equations used in the model are considerably more complex than the Lanchester "Laws" that are often used at the campaign level to model attrition.

One simplification that has *not* yet been undertaken is the introduction of a data-base management system. As noted earlier, a complete run of TAC WARRIOR, generally involving a TAC TURNER run as well, involves about 110,000 equations and takes up to two hours of CPU time for a 10-day campaign. While set-up time for a run may be less than one day, significant changes in the data base can take up to six man-months! It would seem, therefore, that a computerized data management system, and an "executive routine" for integrating it with the simulation, would have a long-run payoff.

OUTPUT

The sample printed output from TAC WARRIOR given in the Users Manual[5] runs to 220 pages, of which 28 pages are a repetition of the input. The user can eliminate portions of the output that are not of interest, stop the run at less than ten days (air superiority may be decided in the first two or three days), call for states of particular interest, and specify the periods for which the reports shall be given. The sample output shows hourly reports (i.e., a read-out every sixty minutes in the daytime, even though the model calculates 1,440 one-minute intervals each day).

The outputs are arranged as follows:

- *Reports of the Stochastic Probability of Kill*
  —Raw Input Report
  —Kill Rates Computation Report
- *Simulation Reports*
  —Summary of Input Parameters
  —Status Report by Time of Day
  —Cumulative Engagement/Kill Information for Individual Aircraft Types for Each Day of the War
  —Executive Summary

The raw input report includes the detection and kill probabilities from the air-to-air and air-to-ground effectiveness models, and the surface-to-air attrition models. The kill rates computation report is the resulting fractional kill rates for each of the possible categories of engagements between loops.

In the simulation reports, the summary of input parameters includes the direct enemy and friendly force information and the states in the airfield portion of the loop. The status report by time of day gives the levels of resources in the various states of all the loops by day and time. As noted earlier, the user can specify the increments of elapsed time or selected times to be reported. The cumulative engagement/kill information reports, for each type of aircraft on each side, the numbers of duels and the numbers of survivors and killed aircraft.

The executive summary is in four parts. The first is a "summary of cumulative CAS sorties flown, AFVS (Armored Fighting Vehicles) destroyed, and armored divisions attacked, by day." The second portion is a "summary output for end of each day" and includes, for each loop and each day, the total number of aircraft surviving, the number killed, cumulative totals killed, and the number of sorties, by day and cumulatively. Third, the "ratio of maintenance support resources remaining to aircraft," is given by loop number. Finally, the numbers of kills of aircraft in air-to-air, ground-to-air, and air-to-ground engagements are summarized.

VALIDATION

It is probably not practicable to attempt to validate TAC WARRIOR by simulating past air campaigns. It is not certain whether sufficient detail on all the elements of an actual campaign would be available as inputs. If they were, the effort might well require more resources then could be made available. It has been suggested that the Battle of Britain might make an interesting validation example, but the effort has not been funded.

On the other hand, it can be said that the structuring of the model in loops of states and transitions between the states provides a considerable amount of "transparency" that permits the examination of what the model has done under particular circumstances, and a good deal of time has been spent in having experienced pilots review events in the air-to-air engagements and air base activities that are at the heart of the model. In the opinion of the principal analysts involved, the confirmation of the reasonableness of the air battles in detailed reviews with experienced pilots have been satisfyingly frequent.

Moreover, there is a certain face validity in the basic TAC WARRIOR logic of viewing the theater air warfare system as a complex logistics and maintenance cycle. The result of each day's activities directly affects the ability of the system to respond to requirements for combat sorties on the following day. Combat losses, enemy interdiction of airfield facilities as well as aircraft and other defense systems on the ground, and down times for maintenance all enter into the trend in sortie-generation capabilities.

Nevertheless, four years of experience with TAC WARRIOR have shown that important elements of a modern tactical air campaign are still not simulated, that many of the simulation steps are based on highly subjective evaluations, and that more work is needed.

Command and Control

It has long been recognized that command and control may be the most important element of all in tac air, particularly as the rapid evolution of communications and near-real time computation takes place. This has led to the requirements for the development of the TAC ASSESSOR model, nearing completion of the first stage of development at time of writing (Fall 1979). TAC ASSESSOR will be more detailed than TAC WARRIOR in command-control, offensive air support, reconnaissance, defense suppression, and ground-battle processes in a two-sided theater level air *and ground* conflict analysis, but will not include air defenses or deep strikes. The first stage of TAC ASSESSOR development has been devoted to the modeling of command and control at corps/TACC and below. The modelers are attempting explicit representation of command and control entities and processes, i.e., the cognitive, planning, communication, and direction processes of command elements, simulated in detail. The command elements observe the battle by requesting intelligence information. They then assimilate that information, use it to "understand" the battle, devise and test plans, and issue plans and directives to subordinate units. Subordinate units behave in a similar manner and incorporate the plans and directives of higher echelons in their higher cognitive and planning processes. At the lowest level of the command entity chain are the actual warfighting elements. These elements are brought to engagement by the command elements. Engagements among these war-fighting elements are simulated, using conventional combat simulations techniques.

TAC ASSESSOR makes extensive use of interactive graphics to permit following the processes occuring in the model. It is also possible to modify the cognitive processes of the command elements during a model run. In this way, the simulated command elements may be interactively "taught" by the console operator and adjusted so that their battle-management behavior parallels that of their "real-world" counterparts.

Explicit representation of the command and control elements is essential to the understanding of the complex interactions expected to occur among various command and control-associated systems that are already in existence or being planned. TAC ASSESSOR is expected to provide, for the first time, a systematic approach to the study of those interactions.

Few-on-Few, Many-on-Many

Similarly, less than complete satisfaction with the simplification of the few-on-few and many-on-many aspects of the air battles in TAC WARRIOR has led to the requirement for TAC BRAWLER. This is a Monte Carlo computer simulation of multiple-aircraft air combat. Each simulated pilot "owns" his own mental models in which he may observe changes in his environment and exchange message traffic with other members of his flight. The primary inputs to the mental model are from simulated visual and radar observations. Each pilot's decision as to what course of action (maneuver) to perform is made using a technique called "value-driven decision making." This technique allows the pilot to consider numerous options for his next maneuver, predict the consequences of employing that maneuver for the near term, weight the results of such a maneuver, and then select the maneuver with the highest score.

TAC BRAWLER is still in the experimental/developmental growth stage. It evolved from the TAC FLIGHT model originally developed by General Research Corporation, and the present contractor is Decision-Science Applications, Inc. It is intended for TAC BRAWLER to replace TAC AVENGER for all of the air-to-air inputs to TAC WARRIOR. Its use should increase the realism of the theater-level models and contribute in particular to the still difficult, indeed unresolved, problem of evaluation of quality versus quantity in fighter aircraft.[11] It appears, therefore, to be an important methodological advance, whether or not TAC WARRIOR proves significantly sensitive to new inputs from TAC BRAWLER.

Electronic Warfare

TAC WARRIOR takes some account of ECM in the surface-to-air attrition models (P001 and the TAC ZINGERs), by assuming that the attacking aircraft turn on their jammers when in range of surface AAA or SAM radars. This is far from an explicit or adequate treatment of ECM (and ECCM—electronic counter-countermeasures), and a new model, TAC SUPPRESSOR, is being developed for the surface-to-air attrition inputs. This simulation will aggregate the systems covered by the TAC ZINGER and P001 models, thus capturing the synergism of multiple Soviet systems (as when a medium- or high-altitude SAM forces an attacker down to a level where he can be reached by AAA or a lower-altitude SAM). It will treat both lethal and non-lethal (countermeasure) means of defense suppression. For each system modeled (an aircraft, a SAM complex, etc.) it will include four elements:

—a decision group—the pilot, in the case of an aircraft;
—a communications group;
—a sensor group;
—and a weapons group (e.g., AIM-7s, AIM-9s, and guns on an F-15).

An interesting feature is that the staffs, or decision groups, can learn and change their doctrines during a ten-day campaign.

TAC SUPPRESSOR will be a Monte Carlo model, too slow in operation to be on-line with TAC WARRIOR but giving outputs that can be used as inputs to TAC WARRIOR. The model is being developed for Air Force Studies and Analysis, with Calspan doing the software.

Density, Rare Events

One general limitation of TAC WARRIOR stems from its basic approach. Since it involves thousands of differential flow equations, the essential variables are treated as continuous. This is suitable when there are large numbers of most of the elements involved, with many encounters and other events taking place in relatively short time periods. Fractional aircraft or SAMs, for example, will not be important, and, in fact, the simulation has a "clean-up" routine which periodically rounds-off to integers. Put another way, the model is suitable for use in a dense environment, as for example in the European theater, where there would be thousands of aircraft and other systems in at most a few hundred-thousand square miles. It is not clear that the model would be useful for a campaign in, say, the Persian Gulf, where there might be say four-percent as many aircraft in an area eight times as large, i.e., about 1/200 of the density of the combat in Europe.

Somewhat related to the above is the fact that the model does not seem to be useful in handling resources that are unique or few in numbers, as has already been observed in the case of AWACS, where there will be at most two or three in the air on each side at any one point in time. If the AWACS is in fact vulnerable, the question of whether it is killed or not killed in any given time interval would have a tremendous, perhaps decisive, effect on the air battle, unlike the killing of any given combat aircraft or ground radar.

Overall Validity

In sum, an evaluation of TAC WARRIOR would have to conclude that the model has, for its time, a reasonable level of in-principle, or relative validity, but it is not sufficiently complete to offer useful absolute findings on the outcomes of a real theater air campaign. Hierarchical itself, it fits into the hierarchy of Air Force Studies and Analysis models and is potentially useful in combination with differential-equation theater ground battle models.

TAC WARRIOR has already proved useful in the achievement of its basic purpose of comparing different aircraft and force structures, as well as some alternative employment concepts for given force structures.

CONCLUSION

It is believed that this review of TAC WARRIOR has illustrated the tremendous complexity of tactical air operations, both by the large number of interacting missions as well as weapons systems and other elements that it has incorporated and by the important elements (e.g., command and control) and complexity (e.g., full and explicit treatment of many-on-many air battles and of pilot decisionmaking in air battles) that it has omitted. TAC WARRIOR has shown a growth capability, having gone through five major versions since its initial development. It has grown extensively, adding ground sectors (there are now twelve sectors on each side), and intensively, adding CAS loops for better resolution and improving methodology for air-to-air battle analysis. As an important part of the learning it has provided to the Air Force, one would have to count the definition of the requirements for TAC BRAWLER, TAC ASSESSOR, and TAC SUPPRESSOR, as well as improvement of the micromodels currently used. At the same time it has grown in complexity and may be approaching diminishing returns on further expansion.

If old warriors never die, TAC WARRIOR will live on with inputs from an expanding array of micro models and perhaps also in still-unforeseen approaches to the solution of the overwhelming problems of modeling a theater-level war, as discussed in the next chapter.

Questions

1. Given that command and control will be modeled TAC ASSESSOR, how might the value of command and control in tac air better be measured?
2. Can air campaign models be expected eventually to achieve predictive capabilities?
3. Are there significant additional omissions or defects in TAC WARRIOR, besides those noted above?
4. Can we ever expect to find an analytical solution to the question of quality versus quantity (e.g., the choice or mix of F-15s and F-16s)? If so, how?
5. What are some important trade-offs and synergistic affects between the various types of Tac Air missions (e.g., CAS, interdiction, etc.)? Does TAC WARRIOR adequately handle these trade-offs and synergistic effects?

References

1. Douhet, Giulio, *The Command of the Air*, 2nd Edition, 1927, translated by Dino Ferrari (New York: Coward-McCann, Inc., 1942; reprinted edition, 1972 by Arno Press, Inc.).
2. Keethler, 1Lt. Gergory A., "A Critique of Air-to-Air Simulation," School of Engineering, Air Force Institute of Technology, Wright-Patterson AFB, Ohio, 25 February 1980.
3. *Military Handbook: Survivability/Vulnerability, Aircraft, Non-Nuclear General—Volume 1 (MIL-HNBK-XXX-1)*, Washington, D.C.: Department of Defense, 30 April 1979.
4. Burgin, George H., *et al.*, *An Adaptive Maneuvering Logic Computer Program for the Simulation of One-on-One Air-to-Air Combat, Volume 1: General Description*, Washington, D.C.: National Aeronautics and Space Administration, September 1975 (NASA CR-2582).
5. *TAC WARRIOR User's Manual*, Draft Report, Control Analysis Corporation, Palo Alto, California, February 1978.
6. *Reference Data for Radio Engineers*, Fifth Edition (New York: Howard W. Sams and Company, Inc., subsidiary of International Telephone and Telegraph Corporation), p. 26–13.
7. *E-3A Airborne Warning and Control System Operational Description*, Rev. Ltr. D204-12301-2 (Seattle: The Boeing Company), p. 1–2.
8. "An Appraisal of Models Used in Life Cycle Cost Estimation for USAF Aircraft Systems," Kenneth E. Marks, H. Garrison Massey, and Brent D. Bradley, (Santa Monica: The Rand Corporation, R-2287-AF, October 1978), pp. 23–27.
9. "The Logistics Composite Model: An Overall View," Capt. R. R. Fisher, *et al.*, (Santa Monica: The Rand Corporation, RM-5544-PR, May 1968).

10. *TAC WARRIOR Analysts Manual,* Draft Report, Control Analysis Corporation, Palo Alto, California, February 1978. The Markovian process is described in standard texts, e.g., [3, Chapter 13].
11. *TAC FLIGHT, A Value-Driven Multi-Aircraft Simulation for Analysis of Close Air Combat,* Final Report, Gorman, G. F., General Research Corporation, McLean, Virginia, November 1977.

Bibliography

Introduction to Operations Research, Hillier, Frederick S. and Lieberman, Gerald J., Holden-Day, 1967.

"TAC WARRIOR: A Campaign Level Air Battle Model," 37th Military Operations Research Symposium, June 1976, Briefing by Captain Steve Amdor.

The Simulation of Cooperative Air Combat in TAC BRAWLER, Unsolicited Proposal DSA/P-78-26. Decision-Science Applications, Inc., Arlington, Virginia.

A Command, Control, and Communications Model for TAC BRAWLER, Final Report and Software Documentation, Report No. 53, Kerchner, R. M. and Gorman, G. F., Decision-Science Applications, Inc., Arlington, Virginia, 14 May 1978.

Value-Driven Decision Theory: Application to Combat Simulations, Final Report No. DSA-67, Decision-Science Applications, Inc., Arlington, Virginia, July 1978.

CHAPTER VI

A Theater-Level Model—IDA TACWAR

INTRODUCTION

Theater-level models never should, and rarely try to, represent war in all its complexity. Theater-level models, therefore, cannot *predict* the outcome of a war, a campaign, or a battle. What theater-level models can do is to compare alternative force sizes, force structures, and force mixes, and rank the alternatives according to specified criteria. Theater-level models can also help to assess new concepts and new doctrine for interactions of air forces and ground forces, and interactions among conventional, nuclear, and chemical weapons. All theater models allocate resources to specific missions and areas; thus, they can help assess the quality-vs-quantity issue: is it better to have fewer expensive, high-performance systems or more less expensive, lower-performance systems?

It follows that there is no such thing as an all-purpose theater-level model. Every existing model provides detail on certain activities and suppresses or aggregates others. In most cases, the decision on what activities to emphasize is governed by the mission of the organization that sponsors the model and the uses intended. The model developer, of necessity, must be ruthless in eliminating or aggregating desirable but non-essential activities, if he expects his model to be routinely used. Military models requiring too many inputs take too long to set up for a new run and are rarely used, even when experienced personnel are available. Moreover, personnel turnover makes it difficult to maintain experience.

IDA TACWAR

One of the most comprehensive theater-level model (in 1978) is the IDA Tactical Warfare Model: A Theater-Level Model of Conventional, Nuclear, and Chemical Warfare.[1] It was developed by the Institute for Defense Analyses for the Joint Chiefs of Staff. It is maintained by the Command and Control Technical Center in the Pentagon and is used or planned for use by a number of U.S. Air Force and U.S. Army agencies and commands.

The IDA TACWAR Model is (1978) the latest of a series of theater-level models. The first of the series was ATLAS, a firepower-driven representative of ground combat. The next version simplified the ground combat submodel in order to incorporate more elaborate air combat activities without increasing computer running time. This goal was met; but the ground combat was over-simplified, according to some critics. This criticism was met by increasing the level of detail in the ground combat process, introducing geographic sectors, weapons classes, and other details. By this time, 1970, increased computer memory allowed a much larger computer code to be used without much change in the running time. Taking advantage of this development, the next version, IDAGAM,[2] kept the basic geographic structure, but provided for explicit interaction among ground weapons and increased the number of aircraft types, air munitions, and other details. IDA TACWAR was completed in 1977. The IDAGAM structure was retained; the number of weapon and aircraft types was slightly reduced in order to incorporate nuclear and chemical weapons. Other major changes were made: explicit target acquisition procedures for nuclear and chemical attack, casualties to local civilian populations, and escalation control procedures (see below).

The IDA TACWAR model was sponsored by the Joint Chiefs of Staff. The model was built, therefore, with the objective of providing a balanced representative of U.S. Air Force and U.S. Army activities in a theater. The model contains details not often found in U.S. Army-developed theater models in order to provide target systems for air attack: it also incorporates details not normally found in air models in order to provide for support rendered to air activities by ground forces.

We will use the IDA TACWAR model as a vehicle to describe the features of theater-level models and the problems that a developer or sponsor will encounter in building and using a theater-level model.

STRUCTURE OF IDA TACWAR

IDA TACWAR is a completely automated simulation. When the input data have been completed, the model runs on a computer until the termination criteria have been met. It then stops and prints whatever output has been specified. The running time in a purely conventional war is about one minute for two days of combat. It is interesting to note that two days of combat per minute is also the rate of the Concepts Evaluation Model (CEM), the most frequently used U.S. Army theater model, and of several other theater models.

Elements in Common with Other Theater-Level Models

IDA TACWAR has several major elements in common with all other current theater-level models.

1. *Data Base*

The data base consists of the inventory of forces and weapons, aircraft and airbases, geographical location of terrain types, and co-efficients, such as P_k (probability of kill) to provide a performance measure of an individual weapon or force unit against an individual enemy weapon or force unit. In IDA TACWAR, as in most theater-level simulations, it is relatively quick and easy to change the *numbers* of forces, systems, and weapons for each side. It is more difficult to change the performance of a system and extremely difficult to change interactions among systems because so many variables must be simultaneously considered.

2. *Theater Organization and Terrain*

IDA TACWAR has a coordinate system based on eight sectors that have a variable width and run the entire length of the theater. The forward part of the sector, next to enemy forces, is a battle area which has a uniform and defined depth. In the rear of the battle area are up to three regions, each containing a forward notional airbase and rear notional airbase. Behind these regions is a single area on each side covering the remainder of the theater called the COMMZ (communications zone). The COMMZ is the point of reception of all incoming combat units, aircraft, supplies, and replacement weapons and personnel. It contains an airbase for long-range tactical aircraft.

Each sector contains a fixed number of intervals in depth to represent different types of terrain, barriers, or width variations.

IDA TACWAR, as many other current theater models, forces the flow of ground combat to be one dimensional; ground forces are pistons. It is impossible to surround enemy ground forces. There are many people working on an adequate two-dimensional representation of ground combat, but no solution has yet been found.

3. *Allocation-Assessment Cycles*

IDA TACWAR, as most other existing theater models, operates on a fixed 12-hour cycle for conventional war.† There are two reasons for using a fixed cycle. It simplifies user inputs because all rates (sorties, movement, consumption, attrition) can be calculated for the same time period. And it avoids the need to input two sets of rules to account for daytime and nighttime operations (12-hour cycles are assumed to begin at midnight and noon). A fixed cycle also allows the user to control the running time of the model, which is directly proportional to the number of cycles.

A fixed cycle creates a number of difficulties in accounting for processes and interactions much shorter than the cycle. For example, IDA TACWAR computes the air battle results first, and then the ground battle and close air support. The coefficients must be carefully chosen so that not all CAS mission aircraft destroyed are killed before they can perform their mission.

An alternative to a fixed time cycle is a critical event cycle. Many small-scale (few-on-few) models are driven entirely by critical events. These models require a specified list of critical events and a matching set of internal clocks. After completing an assessment, the model searches these clocks to determine which critical event is next and when it occurs.

4. *Decisions and Allocations*

IDA TACWAR allows the user to choose one of three methods for allocating aircraft to missions: (1) A built-in optimization algorithm to maximize (approximately) damage to the enemy, taking into account the previous allocation, loss rates, etc.; (2) Allocation completely specified by the user in advance for each time period; or (3) A combination of (1) and (2), in which certain missions are specified by the user.

Allocation of missions (called "postures" in IDA TACWAR) to ground forces is determined by either user input or from a calculation of an attack threshold in each sector. If neither side is strong enough to attack, a holding posture is said to exist. Reinforcements are allocated in accordance with the military principle that the attacker reinforces strength and the defender reinforces weakness. There is also provision for the user to set the size of the reserve force.

5. *Air Combat*

IDA TACWAR does not represent individual aircraft, airbases, or SAMs. Instead, these elements are grouped into types of identical perfor-

---

†Nuclear and chemical activities use a six-hour assessment cycle.

mance. There are seven types of aircraft, six major missions, a single type of airbase and shelter, and three types of air defenses. Attacks on airbases not only destroy aircraft and shelters, but also degrade the future sortie rate of that base. Air-to-air engagements are calculated from an algorithm that allows for saturation effects. Aircraft-SAM engagements incorporate SAM/AAA-suppression as well as destruction of sites and aircraft. Interdiction attacks can be made on missile sites, divisions in rear areas, and supply depots. Combat air support results are computed in the ground combat model.

The air combat model computes attrition to aircraft from air-to-air combat, airbase attack, and from ground SAMs and AAA.

6. *Ground Combat*

The ground combat representation resolves to divisions. It does not explicitly treat battalions and nondivisional combat units, such as artillery and armored cavalry regiments. Ten types of divisions are possible, specified by their initial strength in personnel, supplies, and the number in each of a specified number of weapon types, usually ten. Successful close air support sorties are treated as an additional weapon type.

The ground combat model computes movement of the FEBA and the losses of people and of each type of weapon.

7. *Resupply, Reinforcement, Replacement*

IDA TACWAR has a single type of supply for all air and ground forces. The theater logistic system, however, is represented by a 100-node, 259-arc network. The nodes are supply depots and airbases. The flow of supplies may be redirected every resupply cycle, which is usually every 3 or 4 days. This detail provides the basis for interdiction attacks by aircraft from both sides. If a combat unit is short of supplies, its effectiveness is degraded. Other theater models have made different choices. Some have divided supplies into different classes and account for each separately but have fewer nodes and arcs. Some consider attacks on choke points, such as bridges, that do not destroy supplies but delay their arrival.

Reinforcement, as in all theater models, is treated by an input schedule of arriving divisions and aircraft which are either allocated immediately to battle areas or are kept in reserve and allocated later.

Two replacement policies are available. One follows U.S. policy, in which individual replacement personnel and weapons are sent to combat units as they are needed and available. A pool of arriving replacements is maintained at the COMMZ. The other reflects Soviet policy, in which divisions continue to fight until their effectiveness falls below a specified value; they are then withdrawn and reserve divisions replace them in combat. If insufficient personnel and weapons are available for individual replacement, U.S. and NATO divisions can also be withdrawn when their effectiveness falls below a specified value.

Unique Elements

IDA TACWAR represents tactical nuclear and chemical weapons as well as conventional weapons. The following additions were made to the conventional model to represent nuclear and chemical warfare.

(a) A method for deciding in what circumstances each side will use nuclear and chemical weapons. A variety of options are available: preplanned decision; response to a worsening tactical environment, including loss of territory and loss of nuclear or chemical capability; and response to the enemy's initial (or increased) use of nuclear/chemical weapons. These options are called "escalation stimuli" for nuclear and "employment stimuli" for chemical warfare.

(b) A method for limiting the use of nuclear/chemical weapons. IDA TACWAR provides for a set of nuclear and chemical states for each stimulus. These states are defined by a list of allowable target types, number and range of targets by type, priority by target type and range, allowable depletion of nuclear weapon reserves, and permissable damage and casualties to civilians.

(c) Target acquisition capabilities for discrete elements are provided by specifying a set of sensors and target types, weather and terrain, sensor platform (penetrating or stand-off aircraft, ground systems). The target acquisition submodel provides the expected number of targets of each type that are required.

(d) A method for allocating nuclear weapons to allowable targets, while meeting the specified constraints. This is accomplished by an iterative scheme, taking into account available delivery systems, preferred and second best weapon-target combinations, and the target priority.

(e) Assessment of damage and casualties from nuclear and chemical strikes. These two submodels are a very large addition to the IDA TACWAR computer code. Nuclear blast damage is calculated by VN (vulnerability) numbers for each class of target, as a function of protection available, yield, weapon CEP, and a target offset distance computed from sensor accuracies and reporting delay times for moving targets. Nuclear casualties are calculated in the same manner, including the time in days after burst that the targeted personnel become ineffective. Three types of chemical agents and three types of dissemination are treated. The size and movement of the lethal chemical cloud is calculated from the weather conditions. The casualties and contamination of equipment are calculated for eight protection categories.

Elements Not Explicitly Included

The number of elements and degree of detail incorporated in the IDA TACWAR model forced the developer to eliminate the explicit treatment of a number of elements that are incorporated in one or more other theater models.

Intelligence is not treated. At least one theater-level model simulates command decisions based on imperfect information. In IDA TACWAR, all allocations are based on perfect information of friendly capabilities and, except in nuclear/chemical war, on perfect information on enemy capabilities.

Command-control-communications (C^3) are not treated explicitly. Some C^3 delays are incorporated by a fixed input delay, but saturation of the communications network and the loss of headquarters are not represented. Thus, C^3 nodes are not targeted and either physically or electronically attacked. Some airborne EW capabilities are implicitly incorporated by adjustment of aircraft kill probabilities by interceptors and SAMs.

Air-ground interactions are limited to combat air support sorties against engaged divisions and interdiction sorties against supplies and reserve divisions. There is no provision for the ground situation to affect air allocations and no provision for ground support of air operations, such as SAM and AAA suppression near the FEBA by artillery weapons.

INPUT DATA

The availability, collectability, and validity of input data constrain the model developer and affect the model's credibility. All developers and users face a dilemma. If the model is designed to use only measurable data, it will neglect aspects of combat that are generally agreed to be critical. If it demands many data items that cannot be measured directly and must be generated by some analytical process, the credibility of the model's results will be attacked. The use of data by any model, therefore, represents a trade-off between the data required to represent an important combat process and availability of such data.

It must be emphasized that *all* input data on combat processes are uncertain. Most models of combat use fixed performance values for aircraft, weapons, and other systems—speed, payload, fuel consumption, etc.—while in reality, the performance of each aircraft of a given type, each weapon, etc., will be slightly different and will vary in response to environmental conditions. In this case, the uncertainty can, in principle, be measured. Most theater-level models also use a standard number of aircraft in an aircraft unit and a standard TOE for ground force units; while, in reality, each will enter combat in a different configuration. It is more difficult to determine whether such uncertainties are significant. Uncertainties in performance data of enemy systems cannot be measured and represent the judgment of the intelligence analysts. Uncertainties in the number of systems in an air or ground unit tend to be greater, and uncertainties in the total number of units in the enemy force are greater still.

Data derived according to some assumed functional form are the most frequently used data in IDA TACWAR, as is usual in theater-level models. For example, the results of a large number of many-on-many air combat engagements over a 12-hour period are treated as a small number (5–10) of simultaneous, very large, many-on-many air combat engagements. The air combat results, therefore, depend not only on input kill probabilities, but also on the sortie rates on both sides and on the assumption that the mean outcome of a number of relatively small-sized air-to-air engagements is the same as the expected outcome of a single, very large engagement.

In general, IDA TACWAR attrition computations, both air and ground, are calculated by binomial-type equations using detection and kill probabilities for air-to-air, and input allocations and kill probabilities for ground weapons. Most of these data are readily available and in widespread use.

Judgmental data of many types are used by IDA TACWAR and other theater-level models. In most cases, such data required by TACWAR have been used by many other models and are readily available. Other judgmental data in TACWAR primarily reflect the opinion of military officers. One example is the fraction of ground attack aircraft that jettison their ordnance when engaged by an enemy interceptor.

Collectability

The best source of performance data, of course, is a very recent war. Wartime data are particularly useful for theater-level models because many processes can be combined into a single value, such as the expected number of kills per sortie of a particular type of aircraft, weapon, and target. Thus, historical command-control, weather, variation in pilot skill, fire control hardware, and attack tactics are automatically taken into account. Unfortunately, there are many pitfalls in using combat data directly. Since World

War II, we have no enemy records, so that the kills are estimates with an unknown uncertainty attached. The tactics used in the war may have varied substantially during the period over which the data are available and may not represent the tactics for which current forces are employed. The enemy environment will certainly be different in another war from that of Vietnam or the Arab-Israeli War of 1973. But, most important of all, most theater-level models are used to investigate the impact of changing the weapons, organizations, and equipment of a force; overtime therefore, data on past wars become increasingly inapplicable.

Large-scale field tests unconnected with training, such as JTF-2 in the late 1960s, are perhaps the next best source of data. If the tests are on a large scale and over unfamiliar terrain, it is possible to approximate a many-on-many situation. Unfortunately, such tests are very costly. A common source of test data are annual training tests, usually with a preset scenario to insure that all essential elements are covered. Data for such tests should be used with caution for two reasons: the goal is training, not realistic combat; and many pressures exist to make the reported data close to the performance goals stated by the agency controlling the training.

Field tests of equipment are an excellent source of one-on-one performance data, if allowance is subsequently made, as in the Joint Munitions Effectiveness Manual (see p. 115), for combat degradation factors.

Most available of all, but perhaps least useful, are manufacturers' performance estimates. One method for *simulating* the performance of future systems, thus acquiring synthetic combat data, is to run a highly detailed model of a few-on-few against various threats with a variety of performance specifications to determine how sensitive some overall figure of merit, such as kills per sortie, is to variations in performance. A detailed model serves as a synthetic data generator for use in a higher-level model such as IDA TACWAR. Many of the IDA TACWAR inputs were obtained in this manner.

Validity

There are three aspects to the problem of validating performance data in a theater-level model, such as IDA TACWAR. The first is the technical validity of the measurements from which the data are obtained. Such performance data are: reliability (abort rates), probability of hit, probability of kill given a hit, CEP, fuel consumption, etc. Validity is determined by statistical sampling techniques. In principle, uncertainties can be reduced in increasing the sample size as long as the environmental conditions are precisely known and can be replicated. In practice, however, this is rarely possible.

The second aspect of validity concerns the transfer of the tactical data from the test conditions to the combat environment that the model represents. The performance of most systems is affected by the skill and fatigue of the operator or pilot. The threat posed by the enemy fire encountered will affect most performance factors. While combat degradation data exist for the Vietnam environment, no such data exist for the higher threat environment of Europe. There are many opinions but no agreement. Perhaps the most critical problem is the impact of the significant air threat in Europe. U.S. armed forces have not experienced a major air threat on land since 1942. IDA TACWAR requires an input specifying the reduction in sortie rate after an airbase is attacked and another input specifying the rate of increase if the airbase is not attacked again. While very detailed models exist, such as LCOM (see Chapter V), that are designed to predict sortie rates as a function of damage and repair capabilities, these models necessarily use theoretical estimates not based on experience.

The third aspect of validation in IDA TACWAR is how performance data are actually used in the model. Performance data in IDA TACWAR, as in most theater-level models, are applied to the average system of each type over a 12-hour period. For example, the model cannot distinguish between two sorties of *N* aircraft and one sortie of 2*N* aircraft. Yet, the former case will generally stress ground and air personnel more than the latter case.

Detection, hit, and kill probabilities in IDA TACWAR cannot be used directly because of the 12-hour time interval. Combat experience and the results of detailed few-on-few models show that the total kills divided by the number of sorties of a given aircraft-weapon combination will be far smaller than the probability of kill. Ground targets may be killed twice if the initial kill was not recognized.† In addition, there may be misidentification of priority targets, and degradation caused by weapon interference.‡

Thus far, we have dealt with the validity of measurable performance data. IDA TACWAR also requires input data that cannot be directly measured. Such data have been developed for all theater-level models and are readily available; but they are frequently criticized and will always be sus-

†In the Korean War, an on-the-spot examination of enemy tanks indicated that a substantial number were killed three times, in the following pattern: the initial attack was with napalm, causing the crew to abandon the tank; an unknown time later, the tank was hit and penetrated by aircraft rockets; still later the tank was hit and penetrated by ground tank or antitank fire. Data are available on the fraction of catastrophic kills of aircraft and ground systems, if it can be verified that all catastrophic kills can be easily recognized hours later as well as immediately.

‡One example of weapon interference is the effect of the burst of one smart weapon on the accuracy of a second smart weapon (or precision-guided munition—PGM).

pect to some extent. The most important data of this type in IDA TACWAR are the fractional allocation of ground weapon types to opposing ground weapons.

IDA TACWAR calculates attrition by conventional weapons from two user input tables. One gives the rate at which each weapon type can kill each opposing weapon type if all of its fire were directed at the opposing weapon type. The other table gives the fraction of the fire of each weapon type that is allocated against each opposing weapon type in a "standard" opposing force. These allocations may differ for attack and defense, and for each side, so that there will be four allocation input tables with up to 100 entries in each. The allocations are proportionally modified during execution to take into account nonstandard initial divisions as well as attrition on both sides during combat.

This scheme provides the basis for output tables that give the source of the kill for all weapons lost on each side. The advantage of such an output is that it assists the user in interpreting the overall results of the simulation. The disadvantage is that the output data are sensitive to the weapons allocation input data. Ground weapons allocation is not a command decision, as in the allocation of air-ground ordnance. Rather, the allocation depends on the terrain and other factors that IDA TACWAR does not explicitly take into account. If these factors were accounted for, the input data requirements for this part of the model would increase by an order of magnitude. The developer made the trade-off that even an incomplete and suspect weapons allocation system is better than none in order to obtain kills by source as an output.

DEVELOPMENT AND OPERATION OF THEATER-LEVEL MODELS

Theater-level models take a long time to develop. Depending on how much history one includes IDA TACWAR took about five years to reach its present form at an average cost of more than five professional man-years per year. Similar effort and time was required for the other major theater-level models in current use: Concepts Evaluation Model (CEM),[3] Lulejian,[4] and VECTOR.[5]

It is almost impossible to determine how much effort and time was expended on any of these models in their present form. Most complex simulations of combat are being constantly modified, sometimes by the original developer, sometimes by other contractors, and sometimes by the government agency responsible for the model's operation. In some cases, the modifications are so extensive that the result can be considered a new model and is given a new name. This was the case of IDA TACWAR as it evolved from IDAGAM. In other cases, the revised version is assigned a number: CEM-IV and VECTOR-2.

Let us assume that an entirely new theater-level model is required because a review of current theater-level models has found none adequate even in principle, for the problem at hand.† Past experience indicates that the following steps, each taking about one year, are necessary.

—Develop overall architecture and design specifications

—Develop or adapt algorithms for individual routines

—Research, adapt, or develop input data requirements

—Program and debug individual routines and major sub-models

—Make first not-trial runs with user input and make major modifications to control input and output

—Modify to incorporate user-directed changes in weapons systems, doctrine, tactics, etc.

This sequence, of course, was not exactly followed in IDA TACWAR or any other model, but it reflects the general pattern that has been observed in model development.

The overall architecture and design specifications are usually in the form of flow charts to guide the programmer. At this state, the architecture is guided by relatively broad and simply stated objectives that, in principle, meet all the sponsor's requirements and, at the same time, make the model fast and easy to operate. This is usually interpreted to mean simple and quick changes in inputs rather than computer running time. Often the sponsor also specifies modularity, i.e., the ability to use more than one set of routines, especially some that have already been developed. This is easy to do at the flow chart level, but is much more difficult to program.

The programming stage often produces several problems. First, the broad compass of theater-level models and their cost usually result in a fairly large number of agencies being represented at the progress meetings. The people at these meetings discover for the first time that the military functions for which they are responsible are not represented in enough detail for the model to be of much use to them. The original sponsoring agency and the developer then face a dilemma. If the criticisms are ignored, they lose the support of that agency. If they try to meet a significant

---

†One sample is the Combined Arms Simulation Model currently under development for the U.S. Air Force.

number of these criticisms, the model quickly becomes unwieldy, difficult to control, and the input requirements escalate in number and complexity. Sometimes, owing to the inexperience of the original sponsor and the developer, these difficulties are not recognized until all the individual routines are put together and the entire model is run for the first time. In one case, the effort to simultaneously meet the requirements of many agencies resulted in a 6-month delay and a model too cumbersome for anyone to use.

The development and debugging of the master program is a longer process than most developers recognize or are willing to predict. IDA TACWAR, CEM, and the other theater-level models contain between 20,000 and 50,000 FORTRAN statements. Besides technical bugs that can take time to discover and fix, most models require "tuning." Early runs of a complex, debugged model often produce an overall pattern of warfare that everyone would consider "unrealistic." The pace of war is too fast; there are too many kills; and there may be unexpected interactions among elements that produce peculiar results. This is particularly true of IDA TACWAR, CEM, and other models that use *threshold* values as a control. The interaction among many thresholds can produce distorted or simply foolish decisions by the simulated commanders. It is frequently difficult and time-consuming to "tune" (adjust) these thresholds so that the overall pattern of war and pace of battle are reasonable. Sometimes, it is necessary to adjust weapons system performance inputs to reflect the level of aggregation in time and space (as described on page 142, above).† This tuning process can take several months to accomplish.

After the model has been completed, tuned, and turned over to the sponsor for use, two other issues arise: input verification and the march of progress.

IDA TACWAR requires about 1000 punch cards of input with roughly ten data items per card. Proofreading input data quickly and easily is an unsolved problem. There are many examples of major inconsistencies in input made by tired and busy people working under short deadlines that have remained undetected for a number of runs.‡

IDA TACWAR, CEM, and many other models provide a preprocessor which displays all input in functional categories so that the operator can check it for typographical errors. What is needed, however, is a preproces-

†Proponents of particular systems are sometimes reluctant to allow input kill probabilities to be changed in a model for fear that some other agency will discover it and use the model value to discredit the program.

‡In one case, the user forgot to provide any ammunition to a portion of the force and, because that portion was expected to be quickly defeated, the lack of ammunition was undetected for some time.

sor that automatically queries the user on the basis of predetermined consistency checks. For example, upper and lower limits to the number of elements in a military unit or at an airbase; displaying all input force structure items that are initially zero; and upper and lower limits to the force size in a unit or a particular geographic location.

Most theater-level models require about three years before they can be run for the record. During that period, many changes will occur in programs, priorities, and knowledge about enemy forces and systems. It is almost certain that a change in the model will be required very quickly to deal with a new program. This begins a process that, in practice, is unending. The result is a constant struggle to keep the program and its documentation up to date. If, as frequently happens, there is a significant personnel turnover in the agency operating the model, the result can critically affect the future of the model. Even if the operating agency manages to maintain a minimum level of documentation, the model slowly becomes larger, more difficult to change, and less responsive to priority demands. IDA TACWAR and CEM are too recent to have reached this point, but past experience tends toward a rough estimate of five years for the practical life of a theater-level model at its originating agency.

Although life cycle costing has rarely been applied to theater-level simulations, a rough cycle consists of three years of research and development, two years of limited production, and five years of production and retrofit.

VALIDATION OF THEATER-LEVEL MODELS

As we have seen, no theater-level model contains all the elements of a theater war. It follows that the historical method of validating a model is a shaky one at best, in the following sense: "We have calibrated our model to the results of the 1973 Arab-Israeli War and we can reproduce its results."

There are two reasons why the historical method should be used with caution. First, the environment and force structure on the two sides may not be typical of those the model is designed to investigate. IDA TACWAR was designed for the environment of West Germany and the structure of U.S. and Soviet ground and air forces. In principal, IDA TACWAR could be applied to the Middle East; in practice, the required modifications would be so extensive that the result would be a new model.

The second reason that the historical method should be used with caution is that a major factor in past war may not be explicitly incorporated in the model. One example is the critical importance (as many believe) in the 1973 Arab-Israeli War of electronic warfare, which IDA TACWAR does not consider.

Types of Validity

There is a large literature on the validation of models in general and theater models of combat in particular. We will structure our discussion of validation in terms of four types of validity:

—Input validity
—Design validity
—Output validity
—Face validity

Input validity is the accuracy, currency, consistency, and authority of the force structure and the system performance data base.

Design validity is the degree to which the logical structure of the model and its algorithms are internally consistent and reflect the dynamics of combat in a reasonable fashion.

Output validity is the degree to which the model's output enables the user to rank alternative inputs in terms of specified criteria. To have output validity, the model's output must be sensitive to input variations that the user intends to make.

Face validity is the willingness of the decision-maker to make decisions based (at least in part) on the model because he believes that it makes sense.

We will treat each type of validity in turn, except input validity, which has already been discussed.

Design Validity

IDA TACWAR and all other theater models have been designed to use multiple measures of performance to determine a single measure of effectiveness. The latter must be based on a specific unambiguous mission of the theater force.

This distinction can be illustrated by professional baseball. A team's mission for the season is to win as many games as possible within legal and ethical constraints. The appropriate measures of *effectiveness* in this case is the total number of games won. There are a number of measures of *performance,* of course, which have an important bearing on the number of games won; for example, team batting and fielding averages, average numbers of runs scored by and against the team.

There are several aspects that can be directly transferred to the problems of a theater commander. First, the team that excels in one of the *performance* measures will not necessarily excel in *effectiveness* (i.e., win the most games). Second, in selecting team members, a manager cannot concentrate solely on any single performance measure, say, batting average. Third, the manner in which a team is employed in a particular game depends on the team mission. For example, during most of the season, the manager is unlikely to make use of his entire pitching staff in an all-out effort to win each game. In the seventh game of a World Series, the team mission would be different, and an all-out effort would be appropriate.†

The unambiguous theater mission for NATO is to preserve the territorial integrity of its members, which, in the case of central Europe, means primarily West Germany. This single measure of effectiveness is the amount of territory lost or gained and is measured by the location of the FEBA.

IDA TACWAR and all other theater models move the FEBA as a function of the local commander's mission, the local force ration, type of force, and terrain. Unfortunately, the functional relationship between historical FEBA movements and force ratios is obscure. Studies of this problem have shown no statistical correlation between FEBA movement and force ration.[7] Some believe that the reason for the lack of correlation is that only a small and unknown portion of the force on either side actually fired weapons and contributed to the FEBA movement. This explanation, however, does not improve matters, since it provides no data for modifying IDA TACWAR to account for non-firing weapons on both sides. Others believe that the force ratio-FEBA movement correlation is obscured because ground forces rarely advance beyond the point assigned by the senior commander or, if they did, it was not reported because a further advance would be a violation of orders.‡ Thus, the reported rates of advance would not correlate with force ratio, especially in the critical case of high force ratios and high rates of advance.

Regardless of the explanation, the validity of the FEBA movement algorithm in IDA TACWAR has been seriously questioned and doubt has been cast on the validity of the entire model as a consequence.

A related problem is the inability of IDA TACWAR and all other current theater models to simulate two-dimensional combat in which one force can envelop or surround another. One cause of delay in solving the two-dimensional ground combat problem on a theater scale is the lack of data on the relative effectiveness of defensive weapons when the force is attacked from the flank or the rear, and this is a basic weakness.

Other omitted elements affect the design validity of IDA TACWAR as well as all other current theater models. The most important of these omit-

---

†These two paragraphs are adapted from reference 6.

‡This point was repeatedly mentioned by Marshal Rommel, the World War II German army commander, in his memoirs, and by a number of senior U.S. commanders in interviews by Philip Lowry.

ted elements is the lack of explicit treatment of command control. Within the command and control process, the most critical is how to degrade the information used in the model for making decisions' and allocations. This omission may be of particular importance if the model is used to evaluate the quantity-quality tradeoff issue.

If the theater model allocates a limited number of high-performance systems on the basis of perfect information, the performance of these systems and their contribution to the overall campaign may be exaggerated. Imperfect information will waste some fraction of these systems on unproductive missions and may, therefore, vitally affect the derived tradeoff between many lower-performance systems and few high-performance systems.

A large technical literature exists that evaluates the merits of various algorithms that are designed to approximate the results of detailed few-on-few simulations of specific systems. Few of these critiques, however, explicitly identify the assumptions on which they are based. For example, the form of the Lanchester equations (or their stochastic analogs) is frequently criticized as either assuming too much or too little local control of weapons. In practice, however, the difference in outcome between various forms of Lanchester equations is trivial when the casualties to the winning side (side with the fewest casualties) are less than 30 percent. Similarly, the use of binomial equations to approximate a many-on-many air-to-air engagement has been criticized as implying too much control. In reality, the difference between random and perfect assignment of interceptors to target aircraft is trivial unless the overall kill probability of the interceptor is more than about 0.3 per pass.

IDA TACWAR is a deterministic model, using the mean outcome of a large number of like engagements as its input data. Its fundamental assumption is that the law of large numbers applies to combat. The variations among individuals and lower level commanders are assumed to be small in comparison with the mean; and human factors (skill, leadership, courage, cowardice) can be treated as statistical means.† Nevertheless, high-level decision errors may be decisive.

Brooks[8] has shown that the law of large numbers (termed "stochastic determinism") applies as long as the non-zero correlation coefficients measuring the interactive performance of weapons on the same side are few enough and weak enough. By "interactive performance" is meant the impact of the presence or absence of one weapon on the performance of another. For example, the Israelis found, in 1973, that the effectiveness of

†It should be pointed out that all military planning and procurement decisions in peacetime are similarly based on *average* human factors assumed to exist in the future war for which the decisions are made.

their tanks was degraded by the absence of artillery. The design validity of IDA TACWAR, and of many other theater models, can be questioned if either side has an extremely unbalanced set of ground or air systems. However, it would be difficult to evaluate tactical interactions for weapon systems that have never been used in combat and so define quantitatively a "balanced set of systems."

Design validity can be discussed, as we have, at many levels of detail. Consistency, however, may be a more important criterion for judging design validity than calibration of the model to limited and uncertain data and past combat actions. Unfortunately, there is little literature available on consistency in theater-level models. The major issue is consistency between those combat processes that are treated in detail and those that are not.

One solution is the use of the so-called hierarchy of models. A low-level simulation of combat develops the interactions of a limited set of friendly and enemy systems in one or more engagements with a duration of a few minutes. The results of low-level simulation are aggregated for use as inputs to a high-level simulation, lasting perhaps 24 hours. These results, in turn, are aggregated for use as inputs to a theater-level model. Some have proposed that the hierarchy be an on-line process while running a theater simulation.[9]

The problem raised by the heirarchy solution is identical to the original cause of the consistency problem. The simulation becomes too unwieldy and complex to be routinely used. Each initial set of forces and environment established by the theater-level model would have to be processed by the lower-level models to obtain outputs for use by the theater-level model.

Consistency, at least for the near term, will probably be interpreted to mean that the processes to be omitted or grossly aggregated must have a reasonably small interaction with those of greatest concern to the user of the model.

Output Validity

Output validity is the sensitivity of the model to variations in the input data. Experience with current theater models, such as IDA TACWAR, has shown that the sponsor and user must be continually on the alert to prevent his model being used to evaluate the difference among competing weapons that are designed to carry out identical missions and whose costs and assumed quantities differ by less than a factor of two. Examples are the M-60 vs. XM-1 tank, and the laser designation by the weapon-carrier aircraft vs. laser designation by another aircraft.

IDA TACWAR results are a function of so many inputs that it is extremely difficult to detect the effect of small differences among weapon systems on overall theater results. All that this statement implies is that theater outcomes are (and properly so) the result of many factors and are rarely determined by a single weapon system. Thus, a theater model should not be used to measure the effectiveness of a single weapon.

A theater-level model, such as IDA TACWAR, can be used to establish the initial conditions (or a set of initial conditions) for a detailed model in evaluating the merits of competing systems.

IDA TACWAR can be sensitive to major changes in force structure, i.e., changing the quantity of a type system by factors of two or so. Even then there can be surprises. Other models have demonstrated, after serious question, that some so-called major changes have little effect on the overall outcome. Finally, after lengthy checking and reruns, it has been demonstrated to the satisfaction of most of those involved, that the surprising result was in fact correct and easily explained. Moreover, the result was helpful to the decision-maker because it enabled him to make the decision on the basis of factors other than the overall combat effectiveness, since the differences were small.

It follows that the user's alternatives for examination should not be too close together. Because of the large number of interactions, only relatively large differences can be ranked with confidence that one is better or worse than another. It also follows that, despite the time and cost of preparing inputs, only one set of alternatives should be changed at a time. Otherwise, it will be difficult to rank them and understand the reasons for the ranking.

Face Validity

Face validity refers to the acceptability of the results of the model by a high-level decision-maker. One way to increase the acceptability of the models is to demonstrate that the overall pattern of the simulated conflict "makes sense."† It corresponds to the perception that every senior military officer has of the campaign under consideration. In some cases, the correspondence occurs naturally. In others, it does not and the user has two choices. He can "retune" the model to correspond to high-level perceptions, or he can try to convince his senior officer that the model is correct and the perception is wrong.

One requirement for changing the mind of a senior officer is transparency. Transparency is the absence of so-called "black boxes" in the model

†There is always the danger that the model makes sense to a particular decision-maker or institution because it seems to confirm particular biases.

that need lengthy technical explanations and a considerable mathematical background to be understood. This is the main reason useful theater models, such as IDA TACWAR, have most of the major inputs in the form of look-up tables with simple interpolation algorithms. These tables, or examples of them, can be displayed and their sources and rationale explained in a few minutes. The senior officer can then make his own judgments of their applicability and validity.

The second aspect of transparency is unique to the automated simulation. Manual war games and man-machine interactive simulations are much simpler to design and program because the decisions are made by human players. They are very useful for training but not useful for evaluation. They are not useful for evaluation because, as one officer explained to the author, no chief of staff wants the basis for a program to rest on the combat decisions of the simulated high-level commander made by a relatively junior officer who may never have served on the operations staff of a senior command. An automated simulation has all of the combat processes built-in. The inputs can be reviewed thoroughly before the runs of the model and modified as necessary. Moreover, IDA TACWAR and all other current theater-level models can be rerun with identical inputs and will produce identical outputs if there are no machine errors. This is impossible for manual war games and man-machine interactive simulations.

PROSPECTS FOR THEATER-LEVEL MODELS

The use of theater-level models has been steadily increasing in recent years. Theater-level models, despite their obvious and not-so-obvious limitations, are finding increasing acceptance at high levels. Theater-level models are the few alternatives to intuition and organized judgment when examining a large-scale war in Europe. And the basis for intuition and judgment is rapidly disappearing as World War II recedes into the past and it becomes increasingly evident that Vietnam and the Arab-Israeli War in 1973 have only a limited applicability to Europe. Payne[10] has emphasized that current models basically derive their data and tactics from World War II. It is not clear that models drawing their principal content from a single historical period can tell whether differences in the weapon composition of forces have pushed us past some threshold that implies a radical change in combat. We need models that combine tactical and technological innovation.

The direction of development of theater-level models is primarily toward command, control, and communications. The C^3 problem is closely related to intelligence, target acquisition, and electronic warfare. The next generation of theater-level models almost certainly will incorporate explicit decision-making (allocations) with incomplete, uncertain, and, perhaps, false information. Some models may provide explicit schemes for allocating intelligence, target acquisition, and electronic warfare resources as a function of the combat situation. Most of the technology involved in these processes is new and has never been in large-scale combat. Our knowledge of enemy tactics and techniques is limited. Thus, performance data as well as methods for simulating these processes pose a major problem.

If the next generation of models can achieve even a limited success, we would expect a significant improvement in our understanding of the interactions of these processes with weapons systems and an improved ability to develop requirements that can increase the effectiveness of our forces.

Questions

1. Can a theater-level model ever hope to solve, or help in solving, the complexities of theater warfare?
2. Can a theater war involving the use—or threat of use—of nuclear weapons properly use the FEBA concept?
3. Substitute chemical for nuclear weapons in Q. 2.
4. Has the air-ground interface problem been adequately addressed in IDA TACWAR?
5. Can C^3I problems be adequately modeled?
6. Can better intelligence on enemy tactical doctrine lead to better theater war modeling?
7. Can theater-level models be usefully adopted to a scenario in which more than one theater is involved? In which strategic strikes also take place?

References

1. *IDA TACNUC Model: Theater-Level Assessment of Conventional and Nuclear Combat, Volume II* (R-211), October 1975, and WSEG 275 (IDA R-211), *The IDA Tactical Warfare Model: A Theater-Level Model of Conventional, Nuclear, and Chemical Warfare*, three volumes, November 1977.
2. IDA R-199, *IDA Ground-Air Model (IDAGAM I)*, 1974.
3. OAD-CR-60, *Conceptual Design for the Army in the Field Alternative Force Evaluation, CONAF Evaluation Model IV*, General Research Corporation, 1974.
4. WSEG 259, *The LULEJIAN-I Theater-Level Model*, 1974.
5. WSEG 251, *VECTOR-1, A Theater Battle Model*, 1974.
6. J. A. Bruner, *Measuring the Effectiveness of Theater Forces*, General Research Corporation, 1974.
7. B. R. McEnany, "Uncertainties and Inadequacies in Theater-Level Combat Analyses," *Proceedings of the 16th Annual U.S. Army Operations Research Symposium*, 1977, Volume 2, pp. 1117ff.
8. F. C. Brooks, "The Stochastic Properties of Large Battle Models," *Operations Research*, 13 (1965) 1.
9. DNA 4335F, *Theater Force Mix Issues*, Defense Nuclear Agency, 1976.

10. R. K. Huber, *et al.*, eds., *Military Strategy and Tactics, Proceedings of a Conference under the aegis of the NATO Science Committee at the War Gaming Centre of the Industrieanlagen-Betriebsgesellschaft, Ottobrunn, Germany, August 26–30, 1974,* Plenum Press, New York, 1975.

Bibliography

WSEG 299, *Comparison and Evaluation of Four Theater-Level Models,* 1976.

R-1526-PR, *Models, Data, and War: A Critique of the Study of Conventional Forces,* Rand Corporation, 1975.

RAC-TP-265, *A Formulation of Ground Combat Missions in Mathematical Form,* Research Analysis Corporation, AD 660328.

G. E. Pugh and J. P. Mayberry, "Theory of Measures of Effectiveness for General Purpose Forces," *Operation Research,* 21 (1973) 867.

CHAPTER VII

Strategic Nuclear Exchange Models

THE PROBLEM

If theater warfare is, as suggested in the last chapter, the most complex problem faced by military analysts to date, strategic nuclear war (also called general nuclear war or central nuclear war) is the most awesome and least understood. Such a war, after all, has never been experienced.

Yet strategic nuclear war has seemed to many to be a relatively simple analytical problem. The number of weapons involved is low because the damage that can be wreaked by each is high. (We must never let ourselves forget that nuclear weapons were invented partly because it was foreseen that they would be highly cost-effective.) The analyst need not deal with millions of men and rifles, tens of thousands of cannons and tanks, and hundreds or thousands of engagements, but, at least initially, with only a few hundred or a few thousand warheads and a few instead of many types of weapons. Hence the two aspects of potential nuclear exchange—offense versus offense and offense versus defense—have seemed comparatively simple to understand and to model. In the first decade of the nuclear age, there was little real analysis undertaken, although a body of theory started to be built almost immediately after Hiroshima and Nagasaki [e.g., 1, 2, 3, 4].

But the game became two-sided after the Soviet detonation of A-bombs and H-bombs in 1949 and 1953, respectively. In practice, the United States still had a near-monopoly, but the Soviet effort to catch up was anticipated (and for a time overestimated), and the mid-1950s saw serious attention paid not only to the possibility of a two-sided nuclear war but also to the problem of offense versus defense. At the time, this meant bombers versus air defense, as in World War II, but it rapidly became clear to all that delivery in the 1960s would also be by means of ICBMs and SLBMs. It was widely believed that defense against ballistic missiles was virtually impossible. The metaphor of the difficulty of "hitting a bullet with a bullet" was widely used (including in a speech, as disinformation, by Krushchev). Some scientists perceived at the time that the large kill radii of nuclear warheads on interceptor missiles made the metaphor a false one, but this perception was not general and the focus of both public and classified debate was on offense only. Victory would come through air (or aerospace) power. The phrase is Seversky's, but the concept is Douhet's.[5,6] It was also widely believed that if Douhetism had not won World War II, it was only because the accuracy/yield combinations of the time were not adequate to the task.

The destructive power of nuclear bombs made it a new war game. The threat of massive retaliation would, it was thought, be enough to prevent war. Then the anticipation of improved accuracy made counterforce—the generals' traditional preference ("the best defense is a good offense")—seem feasible and attractive.

In 1962, Defense Secretary McNamara proposed a counterforce-only, "cities-avoidance," doctrine.[7] The Soviets lost no time in denouncing this as an imperialist plot to make nuclear war possible, there were echoes from some allies and domestic sources, and McNamara soon backed off. There is a good reason to believe, however, that his real reason for shifting was not the political heat but the apparently unlimited demands of counterforce backers—for more warheads, greater accuracy, and so on.[8] A countercity threat appeared to make it possible to put finite limits on the forces, and the concept of "finite deterrence" evolved into "assured destruction." Successive McNamara annual "Posture Statements" defined the expected damage required to constitute an "unacceptable threat," or "assured destruction,"[9] and it was but a short step to the mutual hostage theory of mutual assured destruction (MAD). As Paul Nitze has pointed out, it is more correct to speak of mutually assured destruction, since this requires a corresponding, or mirror-image doctrine on the part of the Soviets.

A doctrine of assured destruction-only for U.S. strategic forces came to dominate the thinking of American "defense intellectuals" and to be the U.S. declaratory policy. It is said that actual SAC targeting has never abjured counterforce, but declaratory policy tends to govern programs proposed by the Executive and funded by Congress. In particular, this policy has inhibited improvements in U.S. missile accuracy, through several specific Congressional actions. It has also tended to dominate U.S. negotiating policies in SALT.[10,11]

Our interest here is not in the evolution of strategic doctrine per se, but in its implications for modeling requirements. It was the growth of ABM technology and system designs that gave rise to a perceived requirement for models that would measure differences in damage to both the Soviet Union and the United States, with different attacks, and different ABM deployments and effectiveness. The ABM debate reached its peak in the Congress (and hence in public) in the 1969 and 1970 hearings and votes on deployment of the Safeguard system. It took place within the Executive branch and the "defense community" throughout the 1960s, however, and most specifically between the 1961 McNamara cancellation of the Nike-Zeus system (because of its presumed vulnerability to saturation by decoys) and the 1967 McNamara announcement of the Nike X Sentinal deployment.[12] (The debate continued bilaterally between the United States and the Soviet Union in SALT from 1969 to 1974, leading to the ABM Treaty of 1972 and its Protocol of 1974.)

In order to analyze the effectiveness of ballistic missile defense (BMD), a figure of merit for the destruction of cities was required, in a period in which declaratory policy and capabilities had ruled out counter-force.† Both population and industrial damage have been included in the definition of assured destruction (of the order of one-fifth to one-third of the Soviet population and one-half to two-thirds of its industrial capacity, in the varying McNamara definition). Population turned out to be easier to estimate, since the Soviets as well as the United States publish census results. (The faith in these published figures and projections of them into the future is reminiscent of the perhaps aprocryphal story of the Director of the U.S. Bureau of the Census at the turn of the century who, when asked how the Bureau estimated population, replied, "We don't estimate, we count them!") Manufacturing capacity—a vastly less homogeneous "population" (in the statistical sense) than the population of people—was estimated in terms of the American concept of manufacturing value added, measured for the Soviet Union by the proxy variable of manufacturing floor space, which was estimated from aerial (later satellite) photography. One could hardly call this a high-confidence measurement procedure, but the defects did not seem to matter, in view of the high correlation between population and manufacturing. (Cities tend to grow up around large manufacturing plants, and factories tend to be located in cities, where labor is concentrated—which came first is a chicken and egg question which does

†When the Soviet build-up toward counterforce capabilities became apparent to all in the 1970s, analysis still tended to focus on potential damage to cities by residual forces after counterforce exchanges.

not need resolution here.) The curves tend to look like those in Figure 31.

The greater concentration of manufacturing than of population is more marked in the Soviet Union than in the United States, primarily because of the much larger rural population of the Soviet Union, though this factor is somewhat offset by the smaller number of very-large cities (manufacturing centers) in the Soviet Union than in the United States.

From the above facts and arguments, it came to pass that the first estimates of the relative effectiveness of different attacks (or the same attacks against different defenses) tended to be based primarily on population casualties. In fact, the measure was generally fatalities, "prompt" fatalities being easier to measure than delayed fatalities from fall-out and the impact

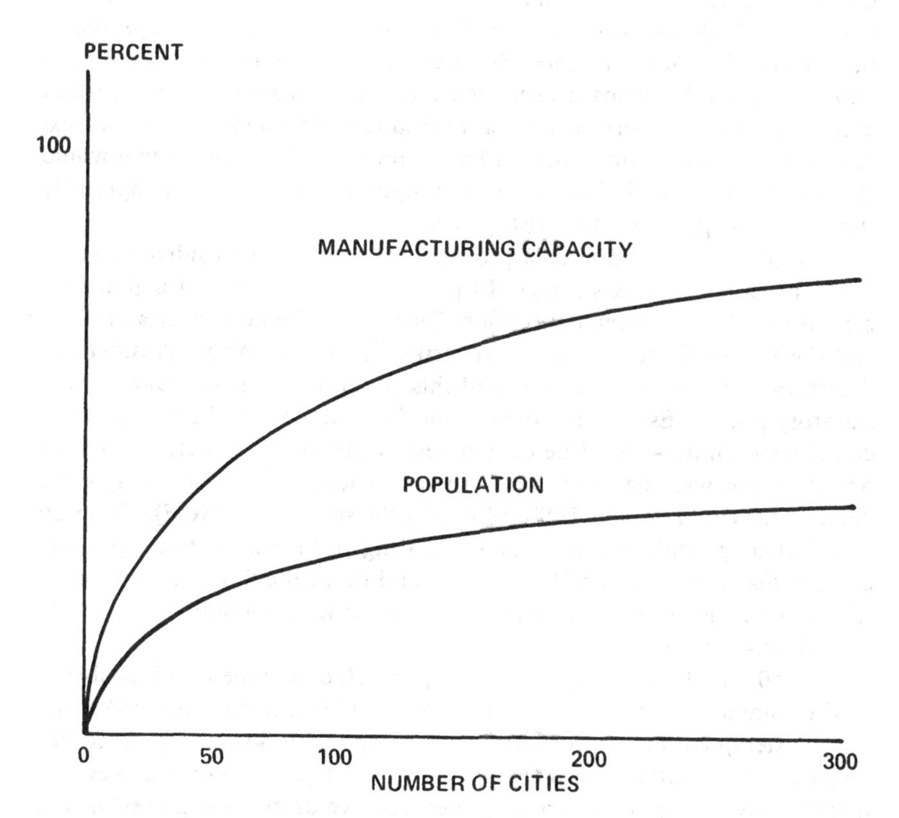

FIGURE 31. Soviet Population and Industrial Capacity, by Number of Cities

of injuries and other non-fatal casualties. This figure of merit, rather grisly, even for war games, was in fact encouraged by the Systems Analysis office in OSD, which was dominated in the 1960s by economists who developed a theory of weapons requirements based on the "demand for fatalities." The phrase, "equal appetites for fatalities in the Soviet Union and the United States," was actually used! As an aside, we may note that it can be argued that the emphasis on the "body counts" in Viet Nam was a result of the same thinking.

CODE 50

One of the early attempts at measuring assured destruction was developed in the mid-1960s and was known as Code 50.† There had been a proliferation of models among the three Services and their contractors, and Secretary of Defense McNamara expressed a desire for standardization so that each party could not bring in different results and then say "but let me explain our different assumptions." McNamara wanted one model that would test the impact of variations in the assumptions. He was only temporarily successful; the problem still exists today.

Conceptually, Code 50 was a program for solving a two-sided zero-sum game, using a minimax strategy. In practice, the assured destruction concept denied the zero-sum assumption. There could be no winner and loser, and the United States would never strike first (although *no* President or Secretary of Defense has ever stated this condition so as to make it a declaratory policy). Essentially, then, Code 50 assumed that the enemy (third countries—China—could be considered in the model structure, but the Soviet Union was the only "real" threat) would strike first, doing what damage he could to the U.S. strategic offensive forces (SOF). Code 50 would then provide a means of calculating optimum or near-optimum damage that the surviving U.S. forces could do to the Soviet urban industrial complex in order to measure the "deterrence capability" of the U.S. second strike forces.

Code 50 was the most sophisticated approach of its time to this problem. In the commonly-used scenario of a Soviet initial counterforce strike, the model attempted, before allocating U.S. weapons to Soviet targets, to optimize the Soviet attack on the U.S. SOF. This step was intended to ensure that the Soviets assumed correct values (relative destructive power) of the

†The model was developed by the Lambda Corporation in conjunction with the Office of the Assistant Secretary of Defense for Systems Analysis (OASD/SA).

U.S. weapons, in order to meet the assumed Soviet criterion of minimizing the damage that the surviving weapons could do to the Soviet Union. The values assigned to the Soviet urban industrial target complexes could be proportional to city populations, or a combination, with subjective weighting, of population and estimated industrial capacity. In the latter case, the weights could be modified to take account of U.S. estimates of Soviet evaluation of particular kinds of industrial facilities.

To calculate the "correct" values of the U.S. weapons, a set of values was originally assigned and then compared with the marginal values implied by the damage the weapons created when they were actually allocated to Soviet targets. If the original and the calculated values differed significantly, the assigned values were adjusted and the process was reiterated until it converged on an acceptably accurate solution. The Lagrangian multiplier technique used in this process will be described later.

Damage Calculations

The following equation was used for the expected value of the damage done by a mixture of weapons allocated to a point target, such as a missile silo:

$$D_j = V_j\,(1 - P_{ij}^{N_{ij}} \ldots P_{mj}^{N_{mj}}) \tag{1}$$

where

V_j = Value of target j

D_j = Value destroyed at target j

P_{ij} = Probability that target j will survive one weapon of type i allocated to it (the single-shot survival capability)

N_{ij} = The number of weapons of type i allocated to target j.

For area targets, e.g., cities, the following equation was used:

$$D_j = V_j[1 - (1 + \sqrt{\sum_{i=1}^{m} K_{ij}^2 N_{ij}})\, e\sqrt{\sum_{i=1}^{m} K_{ij}^2 N_{ij}}] \tag{2}$$

where

K_{ij} = a fitting factor for weapon i against target j

While actual cities vary in shape and value distribution, the equation is correct for large numbers of small weapons optimally targeted against large targets with Gaussian, or normal, value distributions. A value of K_{ij} is computed from a known (assumed) single-shot survival probability for target j attacked by weapon i, from the equation

$$1 - \text{SSKP} = (1 + K_{ij})\, e^{-K},$$

where

$$\text{SSKP} = \text{single-shot kill probability}$$

If there are area defenses, the probability of a weapon reaching its target (i.e., surviving) is reduced. If there are terminal defenses, they are considered subtractive, that is, they destroy weapons allocated to the target until the expected value of the number of weapons (including any decoys) arriving equals the number of reliable defenders at the target, i.e., the defense has been exhausted. The D_js in equations 1 and 2 are then maximized, subject to the constraint that the number of targets attacked is less than or equal to the number of weapons allocated, and that the number of weapons allocated to a target, N_{ij}, is always a non-negative integer.[13]

The Use of Lagrange Multipliers

The nonlinear integer programming problems implied by the above equations cannot be solved by conventional Lagrange multiplier techniques, because of the integer (or discontinuous), non-negativity, and inequality constraints noted above. However, in 1963 two new theorems were published,[14] permitting:

1) Setting the λ of the Lagrangian equations and solving for the constraints at which these λ are optimum. (The λ are in effect shadow prices, or implied values of the weapons assigned. Incidently, hence the name of the Lambda Corporation.) This theorem only assures that if one finds the solution, it is correct; it does not guarantee that a solution can be found. However, the second theorem permitted placing a bound on the solution.

2) A solution can be found in the sense of a strategy subject to the calculated constraint that it is within a specified small epsilon (taken in the model to be approximately 1 percent) of the objective function being maximized.

Using these theorems, actual allocations were computed as follows:

1) An initial, heuristic allocation of weapons to targets was made. The value destroyed by the last weapon of each type allocated was used as an initial estimate of the Lagrange multiplier associated with that weapon.

2) Using this initial set of Lagrange multipliers, an upper bound for total value destroyed was computed. If the destruction obtained by the heuristic initial allocation was close enough to the upper bound, that allocation was used.

3) If the initial allocations were not good enough, new allocations were computed, the Lagrange multipliers adjusted, and new upper bounds obtained until an allocation within about one-percent of the least upper bound was obtained.

The Computer Program

In addition to the above-described allocations and target destruction with the Code 50 computer program, the user could make inputs and parameter calculations as follows:

1) The number of weapons available to be allocated, as a function of:
   a) the number of weapon carriers surviving previous strikes;
   b) the fraction of carriers either not on alert or withheld;
   c) the number of weapons per carrier;
   d) the probability that a weapon will fail in such a way that another weapon can be retargeted, taking its failure into account (assuming reprogramming capability following observed prelaunch or boost-phase weapon failure).

2) The probability that a weapon will penetrate area defenses, as a function of:
   a) the number of area decoys (decoys that can be discriminated only by terminal defenses) carried;
   b) the number of reliable enemy defense units; and
   c) the probability that the weapon will fail in flight after it is too late to retarget for its failure (non-reprogramable reliability).

3) The single-shot kill probability, as a function of:
   a) the yield and accuracy (CEP) of the weapon; and
   b) the hardness of the target (vulnerability number or psi required for destruction).

4) The number of weapons needed to exhaust defenses:
   a) equal to the number of reliable terminal ballistic missile defense (BMD) interceptors: or
   b) equal to the number of reliable surface-to-air missiles (SAMs) that can attack air-to-surface missiles (ASMs) for ASMs on bombers.

The user can also call for additional outputs, including:

1) An annotated presentation of the input data;

2) the number of weapons of each type available to be allocated;

3) the single-shot kill probabilities for each weapon-target combination;

4) the probability of penetrating area defenses, for both missiles and bombers in each strike;

5) a summary of target value destroyed in each strike;

6) a summary of the weapon allocation (lay-down) used in each strike; and

7) the detailed weapon lay-down used for each strike. This includes the weapons used for each target, the destruction, the number of weapons destroyed by terminal defenses, and the Lagrange multiplier associated with each weapon used, which is an estimate of the additional destruction one would attain if he were to use one more weapon that type.

Passing the Torch

As has been seen, Code 50 used the relative values assigned to a defender's weapons in an assured destruction exchange for the purpose of converting the two-sided zero-sum game into two separate resource allocation problems. In the assured destruction scenario, the value of urban industrial targets to the initial attacker is zero, and the value of military targets to the side retaliating is zero. By arbitrarily changing these values (as would be done in the evolving strategic doctrine of the 1970s), the model could evaluate mixed attacks against both military and urban industrial targets. In addition, programming refinements were added to improve the efficiency of Code 50 and to permit better modeling of defenses. However, as noted in earlier chapters, the 1970s also brought greater computer capabilities and new modeling techniques.

THE ARSENAL EXCHANGE MODEL (AEM)

The AEM was also started in the McNamara era, in 1964. But AEM has evolved much further than Code 50 did. Although it still uses numerous linear programming subroutines, it is now a large-scale analytic model using modern computers.

AEM gets away from the assured destruction-only approach of Code 50. It has responded at least in general to the evolution of U.S. strategic doctrine, as President Nixon in 1971 and 1972 added the concepts of flexibility and sufficiency,[15] Schlesinger in 1974 added the possibility of limited, or controlled nuclear war, plus an emphasis on economic targets,[16] and Secretary Rumsfeld in 1977 added that the United States should be able to recover from a nuclear war more rapidly than the Soviet Union.[17] Secretary Brown, although he reinstated an assured destruction criterion (200 of the largest Soviet cities) in 1978, has continued to emphasize flexibility and perceived equivalence. AEM has a capability to consider counterforce (SOF) targets, other military targets, and value targets on both sides, and can consider varied scenarios, e.g.,: first strike counterforce, or counterforce plus OMT; second strike counterforce, counterforce plus OMT, or counterforce plus OMT plus value targets; and third strike countervalue. In the above categorization, value targets can include leadership as well as economic targets.

AEM takes into account many more factors than Code 50. Like Code 50, it allocates, in a second strike, surviving weapons in accordance with target values. It does this, however, in a more sophisticated way—not certainly better, however. In its current version, it uses an off-line model, NUCWAL, devised by the Command-Control Technical Center (CCTC) of the Defense Communications Agency (DCA), to generate its DGZs (designated ground zeroes). Starting with a target data base for each side (of which, more anon), points are assigned to each target on the basis of collective judgments of values. Within a given category, relative values may reflect some measure of relative size or importance. Between classes, finding a common "point" currency in which all values are additive is not so easy. The model allows the user to determine the importance of any given target class, specifying that all members of the class (e.g., silos) shall be targeted, or that all members shall be excluded (e.g., a nonessential industry), or that some percentage of the set shall be targeted after some other priorities have been met. This is begging the question, however, for by that step the user is in fact assigning values relative to all other targets.

NUCWAL inputs include the target locations and values, by class, the vulnerability numbers (i.e., estimated hardness) of the targets, the weapons inventory (including numbers, yields, accuracies, reliabilities, and vulnerability, or probability of arrival, if defenses are specified), and nuclear effects data (to determine optimum heights of burst—HOB).

The model then aggregates targets into complexes, that is, groups so collocated that two or more could be attacked by one weapon of a given yield (which may be nominal or that of a particular weapon in the inventory).

Within each complex, it heuristically selects an initial set of DGZs. This set can be iteratively improved by calculating collateral damage to all targets and estimating total damage expectancy (DE), in points, maximizing this DE within any user-specified constraints, such as minimum damage to one target set, maximum damage (where the number of weapons required would be excessive, in which case a maximum of 95 percent is often specified), or reserve to be withheld. A constraint for MIRV "footprints" can also be introduced.

The AEM then uses the NUCWAL-supplied set of DGZs to optimize the lay-down of the available inventory of weapons, under the specified scenario.

AEM has evolved from a model basically designed in a much earlier computer age. It is not yet as complex and flexible as some of the models we have examined in earlier chapters. Work is still going on to make it interactive with the user and to make it compatable with a number of other models, as in the NUCWAL example. Because we wish in this chapter to illustrate some very basic conceptual difficulties with strategic exchange models, we will not take the time for a more detailed description of the model itself.

RELATIVE FORCE SIZE (RFS)

One more model of the strategic nuclear exchange problem warrants discussion here, both because it is one of the more recent developments and because it has been used in the Department of Defense Annual Reports for Fiscal Year 1979 and 1980 (Figures 32 and 33).[18,19] (The Report for FY 1981 has not appeared at the time of writing, but see page 171, below.) This is the Relative Force Size (RFS) concept developed in the Office of the Secretary of Defense, Program Analysis and Evaluation (PA&E) and documented in.[20] The term "concept" is used advisedly, as RFS is really a figure of merit. It is presently calculated on the basis of a model that will be discussed below. It is an attempt to present an overall comparison of the combinations of different weapons systems on both sides, in their different missions against different target sets. RFS was developed to quantify the capabilities of various arsenals on a single scale (although the vertical scales in Figures 32 and 33 have no values shown, for security classification reasons), with the scale chosen to reflect the effectiveness of strategic weapons systems against given target sets.

RFS is simply the number of times a given arsenal can destroy a given set of targets to a specified destruction level. The user specifies the arsenals,

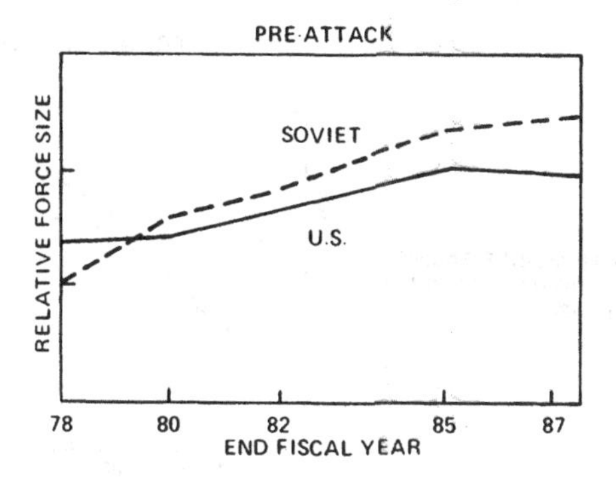

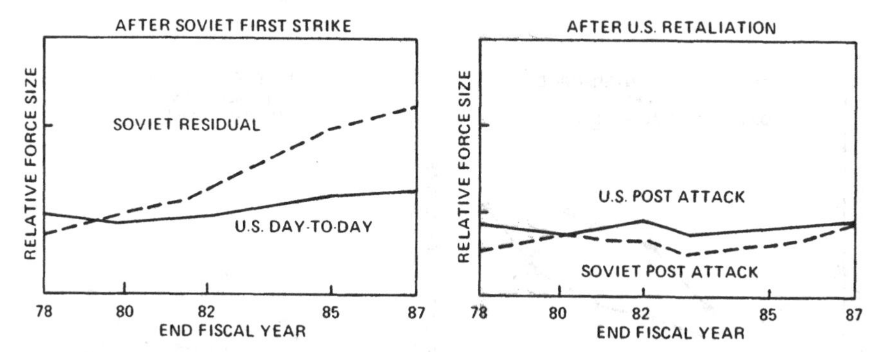

NOTE: RELATIVE FORCES SIZE IS A MEASURE OF FORCE CAPABILITY TO DESTROY A GIVEN SET OF ECONOMIC AND MILITARY TARGETS. THE CHARTS ARE BASED ON U.S. DAY-TO-DAY ALERT FORCES. SOVIET "PRE-ATTACK" FORCES ARE THOSE ON DAY-TO-DAY ALERT; SOVIET RESIDUAL FORCES AFTER A FIRST-STRIKE ARE THOSE WHICH COULD BE GENERATED; AND SOVIET POST-ATTACK FORCES ASSUME THAT THE SAME SOVIET BOMBERS AND SLBMs COULD BE GENERATED PRIOR TO U.S. COUNTERFORCE RETALIATION. WHEN BOTH SIDES ARE ON A GENERATED ALERT, OR WHEN THE U.S. STRIKES FIRST, THE RELATIVE FORCE SIZE MEASURE IS MORE FAVORABLE TO THE UNITED STATES THAN SHOWN IN THESE CHARTS.

FIGURE 32. U.S. and Soviet Strategic Forces Comparison (Day-to-Day Alert)

the target sets, and the destruction goals. The model takes account of the nonhomogeneity of both target and weapons characteristics and seeks a synergistic outcome in which the whole is greater than the sum of the parts. The most important characteristics involved here are those of weapons speed, or time of arrival, yields and accuracy, and, for targets, the degree of hardness, area covered, and time urgency (i.e., mobility—in the extreme case, that of a missile that might have left its silo before an air-breathing but not a missile attacker reached it).

The RFS calculation uses the SOSAC (Son of Super Ace-Central) model for weapons assignment. The allocation technique is quite similar to that used in AEM, in which each column, or strategy, is defined as a resource

ASSUMPTIONS:
- SALT II
- M–X DEPLOYMENT WITH MOBILE BASING
- TRIDENT SUBMARINES WITH C–4 MISSILES
- CRUISE MISSILES ON B–52Gs

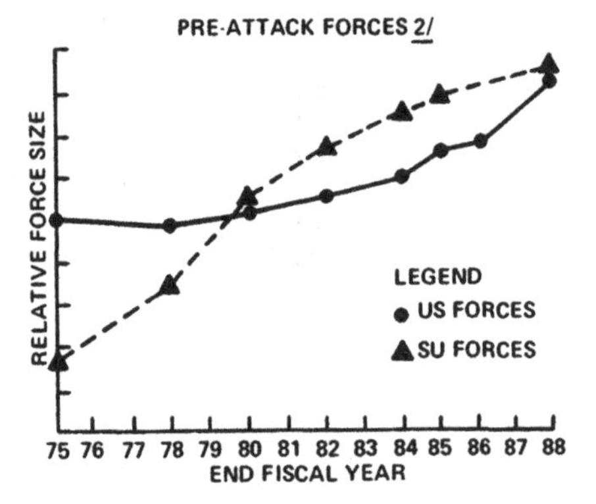

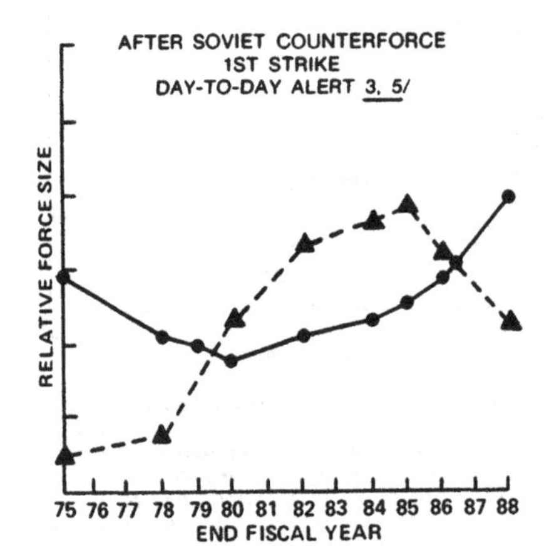

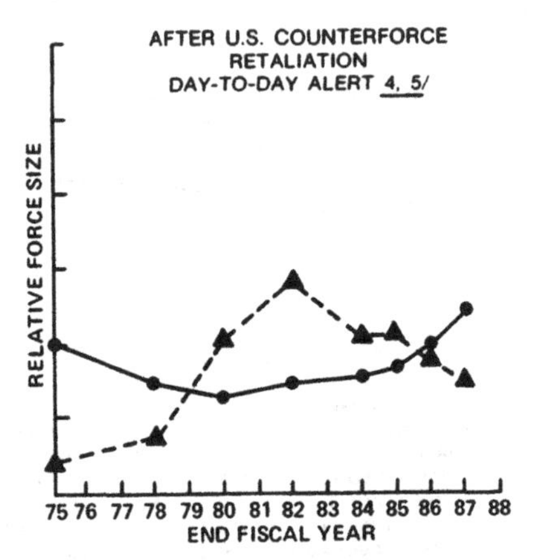

NOTES: THESE CHARTS REFLECT ONLY ONE OF SEVERAL WAYS TO COMPARE FORCES, ALTHOUGH THEY ARE MORE COMPREHENSIVE THAN MOST. THEY DO NOT REFLECT THE BASIS ON WHICH WE PLAN TO USE THE FORCES. AS IS THE CASE WITH ALL MULTI-YEAR FORCE COMPARISONS INVOLVING DIFFERENT FORCES, THEY DO NOT TAKE INTO ACCOUNT CERTAIN OPERATIONAL REFINEMENTS ON EACH SIDE SUCH AS CAPABILITIES OF AND ALLOWANCES FOR THEATER PURPOSES, RANGE LIMITATIONS, AND UNCERTAINTIES ASSOCIATED WITH COMMAND AND CONTROL. IT SHOULD BE EMPHASIZED THAT THE DATA ON SOVIET FORCES BEYOND 1979 ARE SUBJECT TO CONSIDERABLE UNCERTAINTY, BEING PROJECTIONS.

1/ RELATIVE FORCE SIZE IS A MEASURE OF CAPABILITY TO DESTROY A GIVEN SET OF MILITARY AND ECONOMIC TARGETS.

2/ THESE CURVES REPRESENT THE FORCES ON EACH SIDE THAT COULD BE GENERATED (NOT COUNTING UNITS IN OVERHAUL, REPAIR, CONVERSION, OR STORAGE).

3/ THESE CURVES SHOW U.S. DAY-TO-DAY ALERT FORCES THAT HAVE SURVIVED A COUNTERFORCE ATTACK, AND SOVIET RESIDUAL DAY-TO-DAY ALERT FORCES. IF THE U.S. FORCES HAD BEEN ON A GENERATED ALERT PRIOR TO THE ATTACK, THE NUMBER OF U.S. FORCES SURVIVING WOULD BE HIGHER.

4/ THESE CURVES SHOW U.S. DAY-TO-DAY ALERT FORCES THAT REMAIN AFTER A U.S. COUNTERFORCE RETALIATION. SOVIET FORCES INCLUDE SURVIVING ICBMs, ON-STATION SLBMs, ANY ALERT BOMBERS, AND THOSE SLBMs AND BOMBERS THAT THE SOVIETS HAD BEEN ABLE TO GENERATE AFTER THEIR FIRST-STRIKE. IF THE U.S. FORCES HAD BEEN ON A GENERATED ALERT, THE NUMBER OF U.S. FORCES REMAINING AFTER THIS RETALIATION WOULD BE HIGHER.

5/ BOTH SIDES WOULD REMAIN CAPABLE OF ATTACKING A COMPREHENSIVE LIST OF "SOFT" MILITARY AND NON-MILITARY TARGETS AT THIS POINT. FOR THIS REASON, THE HYPOTHETICAL DIFFERENCES BETWEEN THESE FORCES MIGHT OR MIGHT NOT BE MEANINGFUL.

FIGURE 33. U.S. and Soviet Strategic Forces Comparison 1/

package in which a given number of weapons of a single type is used against a target of a single type to achieve a specified level of damage. The allocation process then consists in determining the number of times each strategy must be selected to achieve the specified damage goals. This is basically a linear programming problem, but one that potentially requires an impractically large number of combinations of strategies. The process is simplified by multiplying the number of available weapons in each category by an arbitrary factor and then deriving an LP solution of the allocation problem. If not all the weapon resources are used in the allocation, the factor is too large and must be decreased, and conversely, if the objectives are not met, the factor is too small and must be increased. Iterative corrections then result in the determination of the fraction F of each weapons system in the arsenal which would be required to achieve the desired objectives. RFS is then the reciprocal of F. (RFS could be defined conversely as F, or the fraction of the arsenal required to achieve particular levels of destruction of a given set of targets of different classes.)

The Target Data Bases

The concept is simple. The complex problem lies in aggregating the potential target data. There may be several tens of thousands of identified military and economic (and perhaps leadership) installations in a country the size of the Soviet Union or the United States. Moreover, the targets in such large countries are dynamic, that is, keeping them up-to-date is an almost impossible task that compounds the uncertainties of data collection and interpretation plus the arbitrary task of deciding on the value of different targets for the achievement of the strategic objectives.

The raw target data base consists of the following data about each identified installation:

- category code, or type of installation
- location, expressed in degrees, minutes, and seconds of longitude and latitude
- value, in terms of the worth of the installation relative to the broadly-defined attack objective categories, subject to specified constraints (e.g., "100 percent"—target to destroy all targets in class, say, silo; "75 percent"—target to destroy one-half of total value (points) in a class, say, steel mills; "0 percent"—do not target, say, the textile industry
- vulnerability, usually expressed in terms of static overpressure or dynamic pressure, as a proxy variable for all the nuclear effects that may damage the target

• radius, defining the circle which will totally contain the installation—in practice, for all but "point" targets, the radius containing 95 percent of all the parts of the installation, assuming a circular-normal distribution of target density (a criterion carried over from the early use of "P95" circles for population targets).

Using these data for, say, 30,000 to 50,000 installations, the number of installations is reduced to sets of target complexes roughly ten percent as great in number. These 3,000 to 5,000 target complexes, or aim areas, are then reduced to some 100 to 200 target types, or categories, for computational manageability.

The process of determining the target complexes consists in selecting a nominal large-yield weapon, e.g., one magaton, and then chaining individual installations wherever the vulnerability radii of adjacent targets overlap. The complexes so identified are then attacked with a specified weapon (not necessarily the one used for "complexing"), continuing the attack until some specified level, say twenty percent, of damage is achieved on each installation in the complex. The first weapon is laid down where it returns the most value destroyed. Damage to all installations in the complex is assessed. Those exceeding the twenty percent level assigned to this DGZ and are excluded from affecting the creation of subsequent DGZs. The attack is continued on the basis of the same criterion until all of the installations have received at least twenty percent damage.

At this point, a list of DGZs has been established, but not a weapons allocation. The next step is to test the response, or damage expectancy (DE) of each DGZ, or aim area, to weapons of different yields. For this purpose, four separate yields are used and "perfect" weapons are assumed, i.e., surface-burst, zero CEP, and 1.0 reliability. Each DGZ is attacked with each weapon until at least 90 percent of the installations in the aim area are destroyed, taking account of the weapon radius and the installation vulnerability. The number of weapons of each yield is then stored. In addition, the value in each aim area, by attack objective category, is also stored.

In the interest of saving running time, the aim areas are further aggregated by the user on the basis of their values in each category and the user's judgment as to the importance of the different attack objectives, or target categories. The user retains the actual values of those aim areas that he considers most important and averages the rest. This is in effect a second judgment, since the original values are set on a judgmental basis between different categories. (For any homogeneous set within a category, e.g., steel mills, objective rankings may be feasible, by using, say, the estimated capacities of the mills to assign proportional numbers of points.) The SOSAC model then selects the actual preferred lay-down that measures the total damage, or amount of "over-kill" achieved by the given arsenal of weapons, or, alternatively, the number of weapons required for a given level of damage.

VALIDITY

We have described three out of the large number of models that have been used during the last two decades as attempts to measure potential outcomes of general nuclear war. In the case of RFS, the attempt is to compare force capabilities dynamically as a function of scenario, rather than simply to compare statically the sizes, or inventories, of the forces on the two sides, as is done in a crude way in SALT by use of a single measure of numbers of "launchers" of given types (with the additional refinement in the SALT II Treaty on the "fractionation" of warheads, that is, the number of MIRVed warheads permitted for a given weapon). There is a considerable number of other ways of measuring the size of the arsenal, or weapon inventory, in static terms, e.g., the megatonnage, equivalent megatonnage,† numbers of warheads, hard target kill capability ("prompt" for ballistic missiles and "delayed" for airborne weapons)‡ and so on.

The cases have illustrated some, though far from all, of the complexities of modeling strategic nuclear exchanges that arise despite our introductory comments about the relative simplicity of the problem as compared to the complexities of large battlefields. One complexity not discussed here, is evaluated at length in another model, QUICK (the QUICK-reacting General War Gaming System, developed in the Defense Communications Agency), namely, measuring the value of "cross-targeting," or targeting valuable targets with weapons from different parts of the TRIAD, taking account of the probability of survival of each type of weapon after a first counterforce strike as well as other elements of probable effectiveness, such as yield, reliability and accuracy.

But there are serious conceptual difficulties with all of the models. While these difficulties may not invalidate the use of some models to compare different weapons, or weapons combinations, they call into serious question their frequent comparison of two-sided outcomes.

†Equivalent megatonnage is defined as $EMT = Y^{2/3}$, where Y = yield.
‡Hard target kill capability is generally defined as $HTKP = Y^{2/3}/CEP^2$.

The Data Base Problem

Let us discuss first a very practical difficulty that in the past has been simply ignored and in the recent RFS calculations has been "resolved" in an improper manner. This is what has come to be known as the Blue Data Base Problem. The Red targets, i.e., those in the Soviet Union (ignoring for the sake of argument the possible targets in China and Eastern Europe), are listed in great detail by the Defense Intelligence Agency. There are, of course, great uncertainties and gaps, due to Soviet secrecy and the limitations of intelligence capabilities. Nevertheless, there is a large list of fixed targets with quite precise locations. Unfortunately, the intelligence community is forbidden by law to gather intelligence on the United States. For population data, the Census is used, and this is probably quite adequate. For economic targets, however, the data base is the Census of Manufacturers, which, for reasons of public policy, aggregates many plants in order to maintain the confidentiality of individual records submitted by private corporations. (In some respects the Bureau of the Census is more restrictive than the Defense community in issuing clearances for the use of its data, even though industrial espionage may be less well-funded and sophisticated than international intelligence.) Moreover, plant locations, where given, are often in terms of Zip Code numbers, street names without numbers, corporate headquarters offices rather than actual plant locations, and so on.† For military targets, other than the strategic forces, the situation is not as much better as one might expect. Where Zip Codes are used for industry, APO numbers are sometimes used for military installations. The sizes (number of troops, vehicles, etc.) are not always specified. In short there are no common procedures, rules, and definitions for the collection of data on the Blue and Red economic and military target systems, and there is thus no way to make useful statements about their comparability.‡

In the past, statements of relative outcomes of potential nuclear exchanges have been made without reference to the noncomparability of the target data bases. RFS has taken a different tack, not ignoring the problem, but resolving it by adopting a common target base for both sides, and spe-

†In one sense, accurate locations are not needed, since only the Soviets, not the U.S. analysts will actually target Blue facilities. However, accurate locations are essential to grouping collocated targets for calculation of numbers of DZGs and damage.

‡Where "leadership" or "C3" target sets are considered, there is also noncomparability, but the asymmetry goes the other way. The locations of major U.S. governmental centers and command-control facilities are well-known (presumably to the Soviets also, and hence targetable), whereas Soviet secrecy has successfully prevented U.S. detection of many such points (especially the extensive hardened leadership shelters for wartime).

cifically, the Red target base because it is better understood and more complete than the Blue base.[21] As can be seen in the notes on Figures 32 and 33, the last two Department of Defense *Annual Reports*[18,19] have put it somewhat more obscurely: "Relative Force Size is a measure of force capability to destroy a given set of economic and military targets." It takes a very careful reading to observe that "a given set" means the same set for both sides. This may have been an analytically convenient way out of a difficult problem, a problem that is costly and perhaps bureaucratically impossible to solve, but it is not an analytically proper way of making calculations. Nuclear war is not a skeet shoot, monitored to divide the targets fairly between the contestants. Patently, the targets in the Soviet Union and the United States differ in many ways. For any given category, the numbers, sizes, and often the degree of passive defense—hardness, camouflage, and dispersal—may differ widely.

Intuitively, one might say that on balance the Soviet targets offer more stringent requirements for U.S. weapons than the U.S. targets do for the Soviets.† This would suggest that a comparison using Soviet targets makes the Soviets look relatively poorer or, to put it more bluntly, makes the U.S. look relatively better. In any event, we can't know this without full intelligence, and the use of a single target data base renders RFS meaningless for two-sided comparisons.

Since this chapter was first written (late 1979), the Soviet RFS has been dropped from the DoD *Annual Report*. The U.S. RFS is simply shown, after a Soviet first strike and U.S. retaliation. The lines for U.S. "day-to-day" and "generated alert" states are monotonically increasing from 1981 to 1989, as the number of warheads increases.[22]

---

†As generalizations, one may say of Red and Blue targets:

1) Counterforce: Red has somewhat more numerous and harder, ICBM silos; more SSBNs and SLBMs (not more bases, although the base complexes are more spread out); more, better dispersed, and deeper inland aircraft bases; and vastly more air defense installations;

2) Other military targets (OMT): The Soviets have roughly double the U.S. conventional forces, more dispersed and likely to be "dispersed forward" at the periphery in crisis/wartime (whereas the U.S. forces, other than the small portion stationed overseas, must first be concentrated for deployment overseas by sealift plus airlift);

3) Economic targets: the U.S. economy is twice as large, but the targets in some cases are more concentrated, despite post-World War II trends toward decentralization, and the Soviets have undertaken more passive defense measures (dispersal, concealment, hardening, protection of personnel);

4) Leadership: leadership targets may be more numerous in the highly controlled Soviet society and economy, but, as noted in the previous footnote, better concealed in wartime.

The Problem of Defenses

Code 50, AEM, and numerous other models, not including RFS, have attempted to take full account of defenses in measuring damage, since only missiles that arrive at their targets create damage. Three types of active defenses have been tolerably well modeled. These are, counterforce (which may be viewed as an attempt to shoot down enemy weapons before they get off the ground), air defenses, and ballistic missile defenses. Little has been done about anti-submarine warfare (ASW), because of both the inherent difficulty of the problem and the convenient assumption that attriting submarines is, at least until some unforeseen technological breakthrough, a slow process and a nuclear exchange will be rapid (of which more anon).

Anti-satellite defenses have not been taken into account, for several reasons. First, their effect is on C^3I, capabilities which the analysts have not yet learned how to incorporate adequately in strategic models. (Earlier chapters have commented on attempts to incorporate command control in tactical models and, in some aspects, in the strategic Advanced Penetration Model. There have been some limited models of command control and communications problems per se, but we have not included any case studies of them in this volume—perhaps in the next edition, when more progress has been made.) Secondly, one may be permitted to speculate that as long as the Soviets are reported to have an anti-satellite capability and the U.S. has only an R&D program, it may be politically wise to postpone analysis of the problem. In this connection, it is fair to mention that the United States had an anti-satellite development program underway in 1963, as an offshoot of its ABM program, but it was cancelled by Secretary McNamara.

Passive defenses have been partially covered in the models discussed, but one cannot say adequately. Passive defense may be viewed as the artificial enhancement of characteristics that occur naturally in targets. Thus, every target has some degree of hardness, however trivial, but various measures can be taken to increase hardness. Similarly, all target sets possess some degree of dispersion. This dispersion may be increased and may sometimes be applied to a single target as well as a set. Missiles have some mobility, in that they are transported from factories to launch sites, but their mobility may be greatly increased so that they can move sporadically or even constantly. The models take account of observed dispersion simply by using the intelligence on locations. Rapid dispersion may mean actual mobility (movement in less than the enemy's intelligence cycle leadtime) and mobile targets are generally simply not targeted in nuclear exchange models. (Mobile submarines may be targeted while in port, and similarly, bombers while on the ground.) Hardness is taken into account in the vulnerability calculations on the basis of DNA data from warhead testing (and even some information from the uninstrumented tests at Hiroshima and Nagasaki). These estimates of hardness may be quite soft, however, because of both uncertainties about nuclear effects and uncertainties in the intelligence, but they are built into the models. A special note on hardness is in order. A factory building may have a hardness of say 5–10 psi, and in special cases perhaps more. These would be considered soft targets, but equipment inside them may be as hard as 30–50 psi, and may be further protected to 80 or 100 psi. Thus, much valuable equipment may be salvaged from "destroyed" economic targets. This was, indeed, the experience in World War II bombings. (It has been argued that this bias is "offset" by the failure of models to take account of fire effects.) (See Chapter Appendix.)

Civil defense has sometimes been taken into account in calculating fatalities, by (1) not counting evacuated populations in the urban population estimates used for fatalities estimates and (2) discounting sheltered population (workers) in cities by some assumed vulnerability adjustment. Such calculations are highly arbitrary, however, in view of the great uncertainties in intelligence on the Soviet civil defense programs and on how well they would be carried out in wartime. Moreover, the differences in U.S. and Soviet attitudes toward and definitions of civil defense have stood in the way of careful analysis. As the CIA has pointed out, Soviet civil defense is very broadly conceived. Soviet civil defense puts a top priority on the protection of leadership, then of key workers and preparations for the continuation of their functions (utilities and essential production) in wartime, and finally on the protection of the general population by evacuation from cities in crisis and wartime and protection from fall-out in the rural host areas.[23] In the United States, civil defense is more narrowly conceived as the protection of population (by evacuation and/or sheltering). The protection of leadership and other preparedness measures have been considered a separate function and, until the 1979 creation of the Federal Emergency Management Agency (FEMA), were the responsibility of a different agency from that in charge of civil defense. As we will see below, this difference in U.S. and Soviet approaches has also reflected different estimates of the effectiveness of civil defense by the two powers.

One further type of defense has been much discussed. This is "launch on warning" (LOW), sometimes more sophisticatedly called "launch under attack" (LUA), or "launch on assessment" (LOA). LOW has been treated in models, to the writer's knowledge, only in the sense of treating missile silos

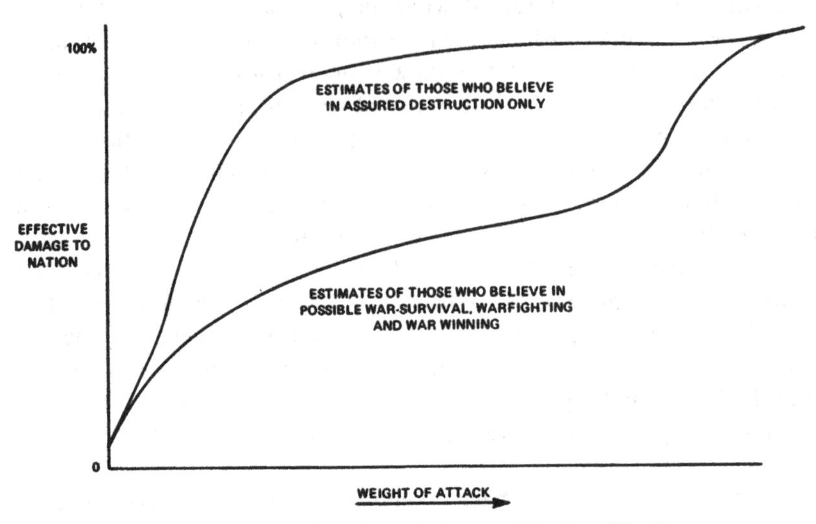

FIGURE 34. Effect of Beliefs on Estimates of Nuclear War Outcomes

as time-urgent targets and by making a distinction in static comparisons between prompt and delayed hard target kill capability. Acquiring a real capability for launch on warning of the U.S. ICBMs† has been discussed as a "quick fix" for the possible U.S. disadvantage in the early 1980s (see Figure 33). The idea has found a curious combination of advocates: those who see LOW as a cheap quick fix to the strength of deterrence and those who see it as a sensible move, if deterrence fails. No one has given an adequate analysis of how the missiles should be targeted or of how the risks of acting on a false alarm can be avoided. For a fuller discussion of LOW, see.[24]

Beliefs and Doctrine

Two polar views tend to dominate thinking about nuclear war. They are represented schematically in Figure 34.[25] The ordinate of the graph is a notional measure of effective damage to a nation as a function of the size of a nuclear attack. One-hundred percent effective damage does not necessarily mean total physical destruction but rather the point at which a society "cracks" and cannot maintain its government, its values, or the capability

†Bombers can always be launched on warning of course, because they are recallable, and so the discussion is not applied to them. SLBMs are generally also omitted, on the assumption that they are not vulnerable to sudden attack (although the idea of launching SLBMs from submarines in port has been discussed).

of independent reconstitution and recovery. The weight of attack, on the abscissa, may be measured in numbers of delivered warheads, megatons, equivalent megatons, or some other proxy variable for all the factors entering into the weight of the attack and reflected in "effective damage."

One school of thought on nuclear war believes in the upper curve in Figure 34. Even small nuclear attacks will have devastating effect (and therefore their threat will deter). The line rises rapidly to the point of virtually complete damage in terms of destroying the national entity, preventing recovery, making the living envy the dead, and so forth. From there on, little is to be gained by the application of additional force—only "overkill," or "bouncing the rubble," in Churchill's vivid phrase. Estimates of the weight of attack necessary to reach this point have varied from one or a few weapons to one weapon on each of the 200 largest cities. (What yield? Should there be a few extra weapons on the largest one or few of these cities?)[18] Accuracy won't matter and enemy defenses will be futile, at most marginally affecting the weight of attack required. Indeed, a precise estimate of this number is not particularly relevant, as escalation control is believed by this school to be impossible once nuclear war starts. Obviously, this line represents the finite or minimum deterrence, assured destruction-only thinking that dominated U.S. policy under McNamara and still dominates to a large extent today.

The lower line would be drawn by those at the other end of the belief spectrum—a minority among U.S. policymakers in the past, although a possibly growing number today, but clearly *the* policymakers in the Soviet Union in the past and presumably future decades. The first one or a few weapons might have similar effects to those projected by the first group (in outcome, but not in deterrence), but then the line rises far less steeply. Defenses, recovery, planning, discipline, and morale can play a role in survival, even in war-winning and recovery. It is only to the far right of the graph that the curve starts to rise toward one-hundred percent effective damage. Even the Soviets appear to agree that there is some very high level of attack that would achieve this result, if only because the contamination by radiation and fall-out would become so widespread as to be insuperable. The weight of attack can be limited by various means that limit damage by moving to the left on the curve. Such means range from SALT negotiations through counterforce capabilities to active defenses, all of which limit the number of weapons arriving on the national soil. Any given point on the curve (short of holocaust) can also be pushed downward by prewar planning, passive defenses, and good wartime organization. None of this is to suggest the abandonment of deterrence as an objective but only that a usable, potentially winning strategy is a better deterrent than a force based on the belief that war is not fightable or winnable. After all, the best deterrent is one that can be used as a threat and may coerce concessions without having to be used. A deterrence-only threat cannot be so used.

The above discussion is not introduced to argue a viewpoint about nuclear war and preparation for it, but to suggest that the kinds of models discussed above, and the figures of merit used in them, are *not* useful when they purport to represent conflict between two powers which tend to play on the disparate curves in Figure 34.

The number of people killed cannot be the correct measure. Does one deter a totalitarian regime by threatening its own victims? Or does it simply provide them with a useful propaganda weapon against the imperialists? Here, it is only fair to note that in 1979 population attacks were dropped from the permissible targeting examples. It was, in fact, discovered that killing noncombatants is illegal in international law. Many have long noted that it is immoral. To complete the Mencken paraphrase, that it is not fattening for military doctrine has been little commented on.

But is economic targeting a better figure of merit? In the first place, large-scale economic targeting with nuclear weapons will kill almost as many people as targeting cities per se. More importantly, war is still essentially conflict between military forces and unless the United States could successfully attack the much larger and more dispersed conventional forces of the Soviet Union, the military value of an economic base to support a long war might be denied the U.S. by surviving Soviet conventional forces, even if the nuclear war turned out to be as brief as in recent decades it has been conventional to assume.

Is Relative Force Size—if it were more validly measured and did not in effect subsume the above figures of merit—really relevant, or is the problem one of attempting to predict the outcomes of various possible war scenarios?† But when we recast the question to one of outcomes, we imply measuring success or failure in achieving objectives. How can this be done between a country which has war objectives (survival and winning in the sense of improving its power position to continue to pursue the prewar aims of the state) with that of a country that has a deterrence-only doctrine without serious damage-limiting survival preparations and war-winning

†To date, attention has focused largely on big attacks—one, two, or three stages in the exchange. But, as Schlesinger pointed out in 1974 and as the emphasis since 1977 on a Secure Reserve Force (SRF) has reflected, there are many reasons to believe that cautious commanders, desperately seeking escalation control, might well use nuclear force in more measured doses and concurrently with conventional conflict on land and sea, so that longer wars of many months, not just hours, days or weeks, are quite possible and might involve nuclear attacks in many stages.

plans? Such doctrine precludes having war aims or a strategy for achieving these aims.

CONCLUSION

Thus far in this volume, it has appeared that, while most military applications of modeling do not yield useful absolute results, their relative findings may be useful for a wide range of problems, from logistics planning to evaluation of weapons and force structure alternatives. In the case of models of strategic nuclear exchanges, however, the models, or at least the results thereof, have come out of the closet of military esoterica, about which the general public couldn't care less, and have been used in official and unofficial debate about national policy. The arguments have been about whether, at various points in time, the forces of the two sides have been of such size and composition as to maintain deterrence in time of crisis, or what has come to be known as "stable" deterrence. The debate has seldom strayed, at least until quite recently, into serious discussion of whether the apparent potential outcomes might become such that the United States would be more deterred than the Soviet Union, i.e., that the Soviet Union might be able to use its strategic nuclear forces for diplomatic coercion of the United States. There has also been little debate over what would happen if deterrence actually did fail. Such an event is generally regarded as either "unthinkable" or as Armageddon (which is the way that many in the 1930s regarded World War II), without questioning whether the Soviets regard it in the same light. Certainly, the models give us little help on this question.

We are forced to the conclusion, then, that to-date the modeling of strategic nuclear war has been a failure, not for lack of technical virtuosity and imagination but because the very first rule of analysis was violated. The problem was not properly stated by a long series of users, and the problem statements were not adequately challenged by a long series of analysts. It should have been said at the outset, by both users and analysts, that you cannot model a war game without a strategy for the war.

Appendix

A NOTE ON INDUSTRIAL HARDNESS

A recent Government report considered both 5 and 10 psi peak overpressure coverage as measures of industrial destruction.[26] Other studies have used a 10-psi criterion. Five psi is sufficient to collapse many structures,

PSI:	5	10	30	50	100
OPTIMUM HOB FOR:					
5 PSI	2.3	.47	–	–	–
10 PSI	2.1	1.0	.03	–	–
30 PSI	1.8	.84	.19	.06	–
50 PSI	1.5	.70	.18	.11	.04
100 PSI	1.3	.62	.17	.11	.06
GROUNDBURST	1.0	.49	.17	.11	.06

(NORMALIZED TO 10 PSI AREA FOR 10 PSI OPTIMUM HOB.)

NOTE: HEIGHT OF BURST OPTIMIZED FOR SELECTED PEAK OVERPRESSURES.

FIGURE 35. Peak Overpressure Areas

and 10 psi will destroy virtually all buildings. But World War II experience and recent experiments have shown conclusively that a great deal of machinery and inventories will survive, partially or wholly operable or usable, under collapsed buildings at 10 psi and greater over-pressures. Hardening for 30 to 50 psi is inherent in much heavy machinery, and various studies have shown that hardening to 80–100 psi or more is feasible for a great deal of equipment, using expedient protection (covering with sandbags, dirt, metal chips from shop scrap, etc.) and longer-term measures (putting key equipment in basements, underground, plus adding redundancy and dispersal in some cases).

A real attack laydown would, of course, not be based on calculations of 5- or 10-psi overpressure radii, nor on heights of burst (HOB) optimized to those criteria. Rather, it would presumably involve optimization of each weapon height of burst to the vulnerability number (VN) of the particular target on the target list. But for targets optimized to higher psi levels, the area covered to a minimum of 5 or 10 psi is reduced. Figure 35 shows the trade-offs involved (calculated from [27]). It shows the relative areas covered to overpressures between 5 to 100 psi when the HOB is optimized to those levels (and for ground burst). For convenience, the areas are normalized to the area for a 10 psi or greater, for HOB optimized to 10 psi.

If the VN (vulnerability) number used for a given target reflects the natural hardness of major equipment, at say 30–50 psi, or of hardening to 50–100 psi, the small area of destruction to this level can be considerably increased by optimizing the HOB accordingly. This will be at the cost, however, of significant reductions in the areas covered to 5 to 10 psi, though the latter areas will still be several times those subject to the higher overpressures. Using the latter criteria will still, therefore, result in considerable overestimation of destruction—and underestimation of assets remaining for potential economic recovery. The majority of U.S. weapons available for economic targeting for the next several years are in the 40- and 200-KT yield classes. Figure 36 (calculated from [27]), shows the radii of the over-pressure circles represented in Figure 35, for 40-KT and 200-KT weapons. If the 40-KT warheads have a CEP of the order of .25 nm, or 1500 feet, they will not give high confidence of imposing 20 to 100 psi on the targets. To achieve 95 percent confidence of a given effect would require a radius of 4950 feet; this is 3.3 times the CEP and is only about the range of 10-psi or less overpressure.

If 200-KT can be air-delivered with a CEP of 600 feet, however, then there can indeed be 95 percent confidence of achieving overpressures of 100 psi. But even in this case, it is not justified to extend the concept of total destruction to the 5- or 10-psi areas.

In short, a great deal of equipment and inventory will survive, particulary in the case of key items for which protective measures may be taken. World War II experience demonstrated, in Germany, Japan, and the Soviet Union, that production can be fairly rapidly restored with surviving

PSI:	5	10	30	50	100
		40KT *			
OPTIMUM HOB FOR:					
5 PSI	7,520	3,420	–	–	–
10 PSI	7,030	5,010	860	–	–
30 PSI	6,840	4,620	2,190	1,160	–
50 PSI	6,160	4,170	2,120	1,680	1,060
100 PSI	5,640	3,930	2,050	1,640	1,200
GROUNDBURST	5,130	3,520	2,020	1,570	1,160
		200KT *			
5 PSI	12,780	5,850	–	–	–
10 PSI	12,020	8,570	1,470	–	–
30 PSI	11,700	7,900	3,740	1,980	–
50 PSI	10,540	7,130	3,630	2,870	1,800
100 PSI	9,650	6,720	3,510	2,800	2,050
GROUNDBURST	8,770	6,020	3,450	2,680	1,980

*40KT AND 200KT WEAPONS (IN FEET)

FIGURE 36. Peak Overpressure Radii

equipment and inventories despite the destruction of buildings. And surviving equipment can be used more intensively in wartime—on more shifts and for more stringently regulated purposes. After a cleanup period—patently, longer and more uncertain than in nonnuclear war—production will indeed be possible with assets from plants that have in many cases been counted as totally destroyed. Equipment may be transferred to surviving buildings, covered with temporary shelters, from tents to Quonset hut-type structures, and even in some cases used in the open (it being not always and everywhere freezing, even in the Soviet Union).

But read on to Chapter VIII.

Questions

1. Can the "poor analyst" effectively influence the definition of the problem and the formulation of the questions for analysis when questions of national policy are involved? If so, how?
2. Can analysis be belief-independent? Must the utility of weapons be "in the eye of the beholder"?
3. Can nuclear exchange models be scenario-independent? Should we seek different models for different scenarios, or sets of scenarios?
4. Is modeling general nuclear war, which has never occurred, a useful undertaking?
5. Can a set of war aims and objectives be defined for nuclear war? Would doing so increase the likelihood of war or the strength of deterrence?
6. Can targeting doctrine be better defined and modeled? If so, could the models be used to aid in decision making about intelligence and weapons design, as well as doctrine?
7. Is it possible to build a model of general nuclear war that reflects the different perceptions (and resultant strategies) of the opposing military/civilian leadership (the problem posed by Figure VII-4)? If so, how would you go about this?

References

1. Brodie, Bernard, ed., *The Absolute Weapon,* (New York: Harcourt, Brace, and Company, 1946), p. 76. See reaffirmation of his views in his last published paper, "The Development of Nuclear Strategy," *International Security,* Volume 2, No. 4, (Spring 1978): 65–83.
2. Kahn, Herman, *On Thermonuclear War,* Princeton University Press, Princeton, New Jersey, 1960, and *Thinking About the Unthinkable,* 1963.
3. Schelling, Thomas, *Strategy of Conflict,* (Harvard University Press), 1960.
4. Wohlstetter, Albert, "The Delicate Balance of Terror," *Foreign Affairs,* Volume XXXVII, (January, 1959).
5. Seversky, Alexander P., *Victory Through Air Power,* (New York: Simon and Schuster, 1942).
6. Giulio Douhet, *The Command of the Air,* 2nd Edition, 1927, translated by Dino Ferrari, (New York: Coward-McCann, Inc., 1942; reprinted edition, 1972 by Arno Press, Inc.).
7. McNamara, Robert, Address to the Graduating Class of the University of Michigan, 16 June 1962, reprinted in The Department of State Bulletin, XLVII (9 July 1962).
8. Enthoven, Alain, C. and Smith, K. Wayne, *How Much is Enough? Shaping the Defense Program,* 1961–1969, Harper and Row, New York, 1971.

9. McNamara, Robert, *Annual Posture Statement of the Secretary of Defense,* January 1965 through January 1968.
10. Schneider, William, Jr., and Hoeber, Francis P., *Arms, Men, and Military Budgets: Issues for the Fiscal Year 1977,* Crane, Russak and Company, New York, 1976, Chapter 2.
11. Nitze, Paul H., "Assuring Strategic Stability in an Era of Detente," *Foreign Affairs,* Volume 54, Number 2, January 1976.
12. McNamara, Robert, Address to the United Press International Editors and Publishers, at the Commonwealth Club in San Francisco, September 18, 1967. Reprinted in *The Essence of Security,* Harper and Row, Publishers, New York, 1968.
13. Lodal, Jan M., "LaGrange Multipliers, Non-Linear Integer Programming, and United States Strategic Force Effectiveness," Strategic Retaliatory Forces Division, Office of the Assistant Secretary of Defense for Systems Analysis, W69-1613, 1965.
14. Everett, Hugh, III, "Generalized LaGrange Multiplier Method for Solving Problems of Optimum Allocation of Resources," *Operations Research,* Volume 11, (1963), No. 3, 399–417.
15. Nixon, Richard M., *U.S. Foreign Policy for the 1970s: The Emerging Structure of Peace,* A Report to the Congress, February 9, 1972.
16. Schlesinger, James R., Speech before the Overseas Press Club, January 12, 1974 and *Annual Defense Department Report FY 1975.*
17. Rumsfeld, Donald H., *Annual Report Department of Defense Fiscal Year 1978,* January 17, 1977.
18. Brown, Harold, *Department of Defense Annual Report, Fiscal Year 1979,* February 2, 1978.
19. Brown, Harold, *Department of Defense Annual Report, Fiscal Year 1980,* January 25, 1979.
20. Battilega, John A. and Grange, Judith K., eds., *The Military Applications of Modeling, Section III,* Science Applications, Inc. (Denver, Colorado, 1978) p. 14.20f, and unpublished materials from and conversations with Captain Gregory Tsoucalis, U.S.A.F. Studies and Analyses.
21. Burke, Lt. Gen. Kelly, DCS/RDA, HQ USAF, paper on "Quantifying the Strategic Balance: Some Common Methods and Their Limitations," delivered at a Seminar on Foreign Policy, Women's Institute for International Relations, Monterey, California, 8 December 1979.
22. Brown, Harold, *Department of Defense Annual Report for Fiscal Year 1981,* January 29, 1980, p. 126.
23. *Soviet Civil Defense,* CIA National Foreign Assessment Center, (Washington, D.C.: July 1978).
24. Hoeber, Francis P. and Amoretta, M., "The Case Against Counter-Force," *Strategic Review,* Fall 1975.
25. Foster, Richard B. and Hoeber, Francis P., "Ideology and Economic Analysis: The Case of Soviet Civil Defense," *Comparative Strategy,* Volume I, Number 4, 1979.
26. "An Analysis of Civil Defense in Nuclear War," U.S. Arms Control and Disarmament Agency, December 1978.
27. Glasstone and Dolan, *The Effects of Nuclear Weapons,* Third Ed., (DoD and DoE, 1977), pp. 105–115.

CHAPTER VIII

Phoenix After the Holocaust: Is Recovery Possible?

INTRODUCTION

Despite the preponderant U.S. emphasis on deterrence, and disbelief in the utility of defense and the feasibility of survival and recovery, described in the last chapter, serious interest in the analysis of economic recovery potential after nuclear war goes back to the late 1950s. By the early 1960s, some of the first large-scale models of recovery were developed. These models were based essentially on the input-output, or interindustry, model conceptualized in the 1920s by Wassily Leontief. Leontief created the first input-output (I-O) table for the United States in 1941.[1] Applications of the technique first became practicable in the 1950s, as computer technology began to be available to handle the onerous calculations involved. Since that time, a considerable number of models have been constructed and a large literature has been developed on the subject.[2]

Before we address the enormous conceptual and analytical difficulties involved in recovery modeling, it is worth noting the key question of whether history (which has had some influence on the modeling) has any relevance in the nuclear age. Historically, economic recovery from disastrous wars has generally proved feasible. The basic reasons for this phenomenon were well stated in 1848 by John Stuart Mill in his *Principles of Political Economy*:[3]

> [The] perpetual consumption and reproduction of capital affords the explanation of what has so often excited wonder, the great rapidity with which countries recover from a state of devastation; the disappearance, in a short time, of all traces of mischiefs done by earthquakes, floods, hurricanes, and the ravages of war. An enemy lays waste a country by fire and sword, and destroys or carries away nearly

all the moveable wealth existing in it; all the inhabitants are ruined, and yet in a few years after, everything is much as it was before. This vis medicatrix naturae has been a subject of sterile astonishment, or has been cited to exemplify the wonderful strength of the principle of saving, which can repair such enormous losses in so brief an interval. There is nothing at all wonderful in the matter. What the enemy has destroyed, would have been destroyed in a little time by the inhabitants themselves; the wealth which they so rapidly reproduce, would have needed to be reproduced and would have been reproduced in any case, and probably in as short a time. Nothing is changed, except that during the reproduction they have not now the advantage of consuming what had been produced previously. The possibility of a rapid repair of their disasters mainly depends on whether the country has been depopulated. If its effective population have not been extirpated at the time, and are not starved afterwards; then, with the same skill and knowledge which they had before, with their land and its permanent improvements undestroyed, and the more durable buildings probably unimpaired, or only partially injured, they have nearly all their requisites for their former amount of production. If there is as much of food left to them, or of valuables to buy food, as enables them by any amount of privation to remain alive and in working condition, they will in a short time have raised as great wealth and as great a capital as before; by the mere continuance of that ordinary amount of exertion which they are accustomed to apply in their occupations.

Feinberg[2] notes that all of the studies bearing on recovery from nuclear war that he reviewed came to the same conclusion as J. S. Mill. This is in part accounted for by the fact that all of the studies assumed the condition postulated by Mill, that the countries "effective population have not been extirpated at the time, and are not starved afterwards." This assumption is justified by the fact that all exchange models of the sort discussed in the previous chapter have shown, at attack levels studied, that more people than industrial plants survive. While people are in general "softer" than equipment, they are also better dispersed, even without urban evacuation (to which we shall return). See, again, the graph on page 174. These findings are, of course, in contradiction to the intuitive belief in the "On the Beach" philosophy, that humans cannot survive a nuclear war, which leads to the belief in the upper curve in Figure 40 and to the convergence of the two curves at the right hand side of the diagram, at "doomsday" levels of attack.

The basic contradiction described here has not to-date been resolved by analysis. Can people survive, not starve or die of fall-out or epidemic, and reconstitute their government, social organization, medium of exchange, and so on, to permit recovery of a modern, complex economy of highly specialized parts? In this chapter we will consider the basic methodologies available for the analysis of economic recovery, and the limitations of these methodologies. We will review in some detail two recent attempts to overcome some of the difficulties and limitations of available methods. In the Validation section, we shall consider the extent to which these limitations have been overcome and the questions about recovery answered.

INPUT-OUTPUT MODELS

Between the two world wars, many economists were preoccupied with finding a way of adequately describing or accounting for the detailed interrelationships of a modern, complex economy involving high specialization, or division of labor, as it had been called in Adam Smith's classic description of the transition to an industrial society. The most successful of these economists was Leontief, referred to above. There was a fine simplicity in his formulation in matrix form of the fact that all industrial sectors or industries take inputs from many other sectors and produce outputs that are in many cases used as inputs by other sectors (intermediate products), not just as products to satisfy final demand. Thus, the steel industry uses coal, iron ore, electric power, refractory brick, etc., etc. But coal mining and transportation, electric power generation, refractory brickmaking, and so on use steel. Many industries, such as steel and coal production, use some of their own output, and all use some labor (the output of the consumer, or household, sector, which utilizes as inputs many end-products from industrial and service sectors). If a square matrix of a sectoral breakdown of the economy is constructed, and if sufficient data are available (e.g., from a Census of Manufactures), "coefficients" can be determined for the consumption of inputs from each sector in producing the outputs of each sector. Some coefficients will, of course, be zero, but surprisingly few in highly-aggregated models, the number increasing with the level of disaggregation.

The concept of the input-output, or interindustry, model is illustrated in a grossly oversimplified manner in Figure 37. In this model of a model, we assume that there is $2 trillion economy that can be described in three sectors: extractive industries, manufacturing, and services. These sectors are used in both the column and row titles. The outputs from the sector in each row are inputs to the sector in each column. In other words, each row

A. DOLLAR VALUES (AT PRODUCER PRICES, IN BILLIONS)

	EXTRACTIVE INDUSTRIES	MANUFACTURING	SERVICES	FINAL DEMAND (GNP)	GROSS OUTPUT
EXTRACTIVE INDUSTRIES	10	60	10	20	100
MANUFACTURING	30	990	400	780	2,200
SERVICES	20	190	90	1,200	1,500
VALUE ADDED	40	960	1,000	2,000	
GROSS INPUT	100	2,200	1,500		

B. INDUSTRY REQUIREMENTS (DIRECT REQUIREMENTS PER DOLLAR OF OUTPUT, IN PRODUCER PRICES)

EXTRACTIVE INDUSTRIES	.100	.027	.007
MANUFACTURING	.300	.450	.267
SERVICES	.200	.086	.060
VALUE ADDED	.400	.437	.666
TOTAL	1.000	1.000	1.000

FIGURE 37. The Input-Output Concept

shows the distribution of the output of a given sector and each column the distribution of the inputs to that sector.

The above description applies to the three-by-three portion of the table at the upper left. The next column to the right adds final demand, which includes personal consumption, gross private domestic investment, net inventory change, net exports and government purchases. The total of this column is the gross national product, or GNP. The final column gives the gross output of each industry, which is the total of intermediate and final demands, and thus includes "double counting." The fourth line of the table is value added, which includes wages, profits, interest, capital consumption allowances, and indirect business taxes. These items are sometimes referred to as charges against GNP and, except for errors and some special items like inventory value adjustments, is equal to the GNP. Part B of the Figure shows the dollar values in Part A converted to coefficients that show the requirements of the industries in each column for inputs from the industries on each line. The inverse of this matrix would show the distribution of the output of the industries on each line among the using industries and final demand, so that the coefficients would add to one at the right column, gross output.

A real world input-output model is therefore far more complex. The most recent published U.S. input-output model is for the year 1972 (a Census of Manufacturers year). The coefficients were published by the Department of Commerce in February of 1979.[4] Note that the input-output coefficients are computed to five significant figures. The dollar values underlying the percentage terms of the input-output model were published in April 1979.[5] The two-month delay in the publication of the dollar values reflected the time required for the "reconciliation of the GNP derived as part of the Input-Output (I-O) study with the GNP derived in terms of incomes as part of the National Income in Products Account's (NIPA's). The tables published in the *Survey* showed 85 industries/commodity categories, or the two-digit level of the Department of Commerce classification. It was announced that the four-digit levels (365 industry/commodity categories) and six-digit levels (496 categories) would be published subsequently but would be available on computer tapes on or about June 1, 1979. These numbers give some idea of the complexity of the enterprise.

The reader will already have noted that it took seven years to collect the data, calculate the input-output relationships, and publish the results. The above numbers may give some indication of the enormous complexity of the task. When it is added that the input requirements for each six-digit industry requires long and detailed study and analysis by specialists, the time lag may become understandable. The significance of the time lag for the usefulness of the results will be discussed below.

The level of aggregation is crucial, for several reasons. First, the greater the disaggregation, or the more columns and lines in the matrix, the greater the difficulty of both constructing and mathematically "inverting" the matrix. The data collection problem for the determination of coefficients rises with the square of the number of sectors (columns or lines), and the number of computations for the mathematical solution, or inversion, of the matrix rises with the cube of the number of sectors. On the other hand, the smaller the number of sectors, the greater the rigidity built into the analysis, since the model permits no substitution between sectors but assumes infinite substitutability within each sector. Thus, if coal, gas, and oil are in separate sectors, they cannot be substituted as fuels in the generation of electricity for the production of steel. On the other hand, if steel is one sector, its product is treated as homogeneous and a unit of steel is the same when used as an input to any sector, regardless of the vast variety in types, shapes, and qualities of steel products.

The reason for the limitations on the substitutability of different materials and products between categories, at whatever level of disaggregation, is apparent if one remembers that, in order to compare input and output "apples and oranges," we were forced to use dollar values as a common denominator. The 496 six-digit categories mentioned above is a high level of disaggregation in terms of analytical manageability and the human capability to comprehend the complexities of the interrelationships. However, a random sampling of some of the six-digit titles is illuminating. Category 17.0100 includes "Floor Coverings"; 27.0100 is "Industrial Inorganic and Organic Chemicals"; 41.0100 is "Screw Machine Products and Bolts, Nuts, Rivets, and Washers." We need not go on. Patently, many six-digit categories include hundreds or thousands of different products, many of which may be substitutable for each other, but many of which will not. We will not know if a rivet is substituted for a bolt and nut, because they are all lumped in one dollar total. Some of the six-digit categories may represent many billion dollars of output—they averaged over $4 billion in 1972!

In the modeling of peacetime economies, changes in coefficients, representing changes in technology, including substitution and other forms of "conservation," take place for the most part relatively slowly (with exceptions for "young" industries such as computers) and can be taken account of in, say, annual or biennial I-O tables—though in fact U.S. tables have been published for 1947, 1958, 1963, 1967, and 1972 only. The 1972 tables were published in 1979, and in those seven years there were dramatic changes in many industries—e.g., computers, typewriters, communications, autos, fuel usage . . . A recovery period, moreover, could be expected to involve changes similar to those of a wartime period, but even more rapid, in that conservation would perforce be very extensive—not over seven years but almost instantaneous—and changes in coefficients therefore very rapid and virtually unpredictable. In particular, one could expect extensive substitution of labor for capital, i.e., greater *relative* value added.

Let us return for a moment to our example in Figure 37. The table shows the (imaginary) input-output coefficients for a $2 trillion economy. In ten years, this economy might become a $3 trillion economy. (The reader will please ignore, with us, the possibility of inflation.) Surely, the coefficients for that $3 trillion economy, reached after ten years of change proceeding at widely-varying rates in different sectors, would look quite different. But imagine, then, changes caused not by the passage of ten years but by a war that included one or more nuclear attacks on the economy in question and in the course of weeks or months reduced the $2 trillion economy to one of $1 trillion, or even $500 billion. Waiving (though only for the sake of argument) the question of whether that smaller economy could be restarted and function, the coefficients of the $2 trillion preattack economy would be irrelevant. Whole industries might have been wiped out. Priorities for production from surviving plants, as well as for construction of new plants, would be entirely different from those of the prosperous $2 trillion prewar economy. Furthermore, the changed relative scarcities and income levels would be reflected in radically altered relative prices (see the "Dynamic Postattack Recovery Model," below).

The input-output table is, then, essentially static. A given set of inputs produces a given output, in effect instantaneously. An important early application of the input-output technique to civil defense planning and economic recovery modeling was PARM (Program Analysis and Resource Management), developed by Marshall K. Wood, *et al.*, at the National Planning Association for the OCDM (Office of Civil and Defense Mobilization) and its successor, OEP (Office of Emergency Planning). This model attempted to modify the static nature of the I-O procedure in one respect by replacing input coefficients with complex records that included lead-time parameters to account for the fact that inputs to an activity are not generally concurrent with the associated output.[6]

Despite its serious limitations, the input-output technique has obvious appeal for the analysis of postattack recovery, because it offers an immediate way of taking account of differences in the damage to different sectors of the economy in a nuclear attack. These differences arise both accidentally and deliberately, in the latter case because some industries may be regarded as more vital targets than others. Indeed, input-output analysis has been used heuristically for determining the relative attractiveness of different economic targets by measuring their impact on other sectors.

Most attractive for an attacker, and of greatest concern for a defender or recovery planner, may be an essential "bottleneck" industry or product. If virtually all of such a category can be destroyed, it is theorized, the economy will grind to a halt, or recovery will be long delayed. Indeed, with zero output of a widely-used product, an input-output model will show exactly such results. In the real world, however, substitution of inputs, plus use of inventories and/or external resources and rationing of outputs, may greatly mitigate the effects of the loss.†

---

†Note that "substitution" has been used in the discussion of input-output models in a very general sense. One input may be substituted for another to create an equal value of output. A new product may be substituted for an old, or for a null set of old products, i.e., it may be a net addition. By the same token, in peacetime and more often in war or recovery times, an input may be eliminated or reduced. Take, for example, machine tools. These have traditionally been mounted on massive iron and steel bases, to give them stability. Over the years, the weight of these bases has been progressively reduced (sometimes by the substitution of lighter rolled steel for heavy iron castings), and during both World War II and Korea, the trend was accelerated in the name of "conservation." Conservation can also include such measures as standardization, producing fewer models of a given item, thus potentially saving labor time, machine time, energy, and materials. Nevertheless, it is convenient to think about changes in input-output coefficients in terms of substitution and substitutability.

Nevertheless, because of the innate appeal of the approach, a considerable number of further attempts to construct input-output recovery models were made in the late 1960s and early 1970s, building on the pioneering work in PARM and taking advantage of growing computer capacity, as for example in the recovery portions of the IDA Civil Defense Economic Model.[7,8] Further refinements were introduced by the application of linear programming, based on the techniques first introduced by George B. Danzig in 1947. In an input-output model, a deterministic solution is derived, in which all surviving output capacity is utilized, in the sense that the output from a given sector is distributed in proportion to the coefficients for that output as inputs to the other sectors, including direct final demand. But, obviously, priorities will be set during the recovery period. Linear programming permits considerable flexibility in testing such priorities, since each output equation is a constraint, that is, it utilizes the inputs from other sectors, but not necessarily completely; e.g., the output of clothing might be constrained to a level below that permitted by the available inputs, whereas the output of steel might be maximized.

ECONOMETRIC MODELS

As we have seen in other applications, the development of much higher power computers permitted (if it was not the driving factor in) the development of simulation techniques by which equations could represent interindustry and other economic relationships in a much more flexible manner, including the use of differential equations, which need not be of the same form for different variables. Technological change, substitution, and policy changes are easily introduced. The inherent difficulties of basing the equations on prewar data and of arbitrariness in allocation policies remain. A recent and ongoing example is the SRI-WEFA model†.[9] This model, called SOVMOD (Soviet Model), is actually a program for the construction of a macro model of the Soviet economy which adapts the econometric model-building techniques developed for Western market economies to analysis of the centrally-planned Soviet economy. It attempts to take account of Soviet budget data and five-year plans, and of the differences in Soviet practices in the determination of labor participation and distribution, investment, capi-

†SRI stands for SRI International, formerly Stanford Research Institute, and WEFA for Warton Economic Forecasting Associates, headed by Lawrence Klein of the Wharton School of the University of Pennsylvania.

tal formation, agricultural production, wages, consumption and foreign trade.

In econometric models of Western market economies, stability, or structural invariance, is assumed on the basis of an approximation to competitive markets; structural stability in a planned economy is, in contrast, based on what has been called "administrative regularity" in the controlled, or command, system. A number of signals have been identified for contingencies that shift important variables significantly away from planned levels. The most important of these in SOVMOD are:

- weather-induced deviation of the agricultural harvest from potential or planned levels,
- defense budget decisions that change the rate-of-growth of real non-personnel defense expenditures,
- changes in the rate-of-growth of real gross profits, induced by current and past harvest variation and changes in consumer goods inventories,
- diviations from the planned rate-of-growth of non-agricultural capital investment, resulting from the changes in harvest and real gross profits noted in the above,
- large changes in hard currency liquidity.

Currently, SOVMOD III is in use for five-year annual forecasts, under varying scenarios. This model uses six sectors, with the industrial sector disaggregated into twelve branches as follows:

1) Industry
   - —coal products
   - —petroleum products
   - —ferrous metallurgy
   - —non-ferrous metallurgy
   - —chemicals and petrochemicals
   - —machine building and metal working
   - —forest products
   - —paper and pulp
   - —construction materials
   - —soft goods
   - —processed foods
2) construction
3) transport and communications
4) domestic trade

5) government and services
6) agriculture
   —grain production
   —total crops
   —animal products
   —meat

Income categories include:

1) Nominal annual wages for seven sectors of employment, corresponding to the above six production sectors, with a division for agriculture between state and collective farms;

2) Household income
   —money wage income
   —agricultural income in kind
   —state transfer payments
3) Nonhousehold income
   —gross profits
   —amortization in state and collective organizations
   —state budget (four revenue categories)
   —expenditures or financing the national economy
   —social-cultural expenditures
   —administration
   —defense
   —residual
   —wholesale prices (two categories)
   —retail prices
   —state sector (two categories)
   —agriculture—a "negotiated" agricultural price for the determination of the consumption price deflator

Foreign trade:

1) By geographical area:
   —Council for Economic Mutual Assistance (CEMA)
      —raw machinery
      —food grain
      —other consumer goods
   —Developed West
      —raw machinery
      —food grain
      —other consumer goods

—other socialist economies
—less developed countries

2) Hard currency liquidity
—current account in hard currency
—credit repayments and interest
—services and transfers
—credit drawings
—gold transactions

Consuption:
—food
—soft goods
—durables
—servies

Production:
—new capital investment
—capital repair
—inventories
—domestic trade
—non-agricultural

The Soviet input-output tables developed by Professor Vladimir Treml (a participant in the project) have been used as an adjunct to the model to supply detailed coefficients of gross output among producing sectors and categories of end-use. The macromodel has been used to feed back information on factor allocation and the composition of output, to indicate changes in input-output coefficients and capacity constraints over time.

SOVMOD is not yet considered usable for projections of economic recovery, although it has been used to test alternative Soviet allocation policies in recovery scenarios. Additional models are being developed which it is hoped will constitute, in conjunction with SOVMOD, a family of models that will be usable to measure potential recovery. These include:

- a medium-term macromodel with an extended foreign trade component,
- a long-term growth model (10–25 years), incorporating technological, structural and goal changes,
- defense-related models elaborating the estimates and specifications of Soviet defense expenditures.

An SRI-WEFA *S*oviet *E*conomic *M*odel of *Rec*overy, SEMREC, is under development. It is intended that the U.S. and Soviet models can eventually be used for comparative recovery estimates.

THE "DYNAMIC POSTATTACK ECONOMIC MODEL"

A model currently under development by Decision-Science Applications, Inc., called the Dynamic Postattack Economic Model, or the DSA Economic Model for short, attempts to overcome some of the limitations of the classes of models described above. The model was developed in response to ACDA (Arms Control and Disarmament Agency) interest in the effectiveness of the Soviet civil defense program and the effect of that program on U.S. deterrence. The contractor has also been concerned with implications for U.S. civil defense measures and for the evaluation of alternative targeting doctrine. The implicit assumption is that the length of the postattack recovery period, if it can be measured, is one relevant indicator of the effectiveness of civil defense programs.[10,11,12]

The analysis starts, as have most others, with recognition that the postattack economic patterns and required allocation policies will perforce be very different from those in the prewar world. Consumption will be severely limited, productive capacities will be greatly curtailed but must be rebuilt selectively, defense requirements may continue or be increased, and import possibilities may be radically altered. As we have seen, input-output and econometric models attempt in various ways to allow for the user imposition of constraints and allocation policies. This process necessarily imposes the user's judgement in the selection of the policies to be tested. The DSA Model, while also permitting user intervention, attempts to establish an objective means for determining new patterns of demand based on the patterns of damage to the economy. (The damage patterns used are exogenous to the model; they may be taken from scenarios run on existing damage models [see Chapter VII] or may be simply assumed.)

The model generates new, and changing, shadow prices to reflect the changed, and continously changing, relative scarcities in the postattack economy. These shadow prices are Lagrangian multipliers, expressed as ratios to their prewar values, all of which are assumed to be 1.0. The final report states that "To understand the relationship between economic priorities in the peacetime economy and those in a postattack economy it is necessary to distinguish carefully between the fluctuating economic priorities and the relatively stable value judgements that determine the basic economic demand curves."[10]

An example of the patterns of the shadow prices as they change over time is given in Figure 38. Note that the price of labor is *down*; workers may be lucky, in fact, to get 50 percent of prewar wages in year one, even with the price of consumables up 700 percent. "To model the behavior of a postattack economy, it seems appropriate to assume that the underlying

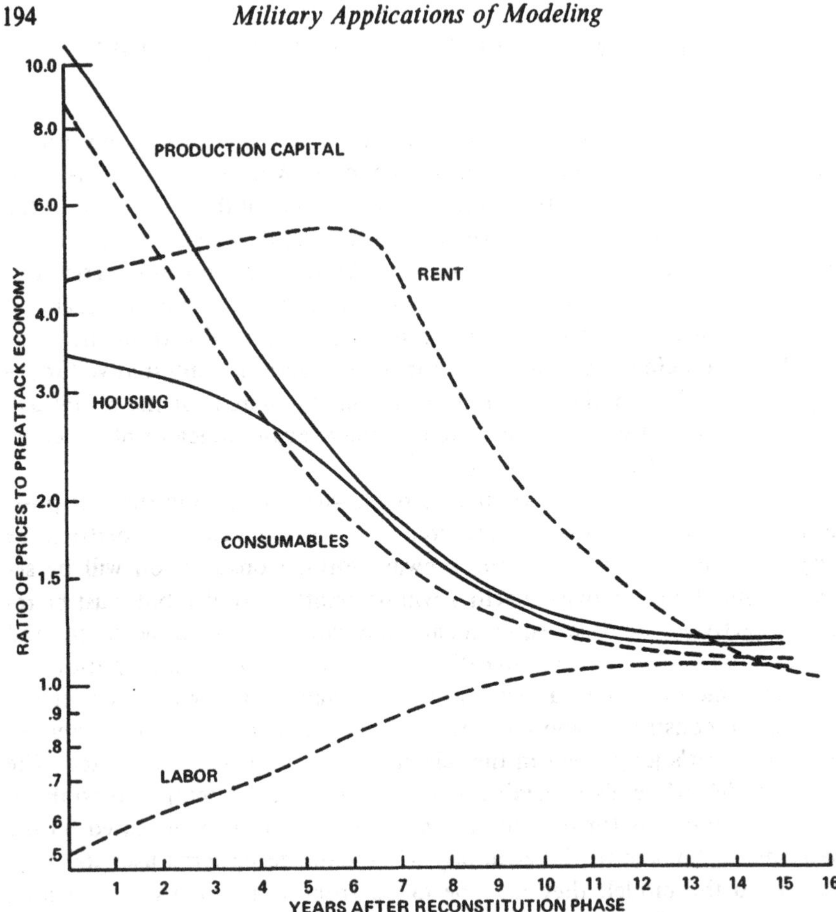

FIGURE 38. Projected Prices in a Postattack Economy (Based on Shadow Values)

value judgements reflected in economic demand curves will remain relatively stable, but that there will be large variations in economic priorities as a consequence of severe shortages that can be expected in the postattack economy."[10] The writer believes there is a strong possibility that the underlying value judgements may be altered in such a world, at least for some years, and that these changes will not in fact be fully reflected in the shadow prices.

Data

For data, DSA, like eveyone else, is stuck with playing "the only game in town": it uses the basic input-output tables published by the Department of

Commerce. (There is a preprocessor used with the model to generate any user-prescribed aggregation of sectors, plus damage levels to these sectors. The preprocessor could, incidentally, be very useful in aggregating to the same number of sectors, as well as can be, the differently-constructed U.S. and Soviet I-O tables.) The model can then take two steps; first, it can generate an equilibrium state that could be reached, starting with the initial postattack (and post survival and reconstitution phases) economy, and second an optimal path over time for reaching this equilibrium, iterating the calculations with successively adjusted Lagrangian multipliers until the stable equilibrium in reached. We will describe the model briefly in terms of the principal equations used. The terms are defined in Figure 39.[12]

Constraint Equations

There are seven constraint equations used in the model that need to be stated before the objectives can be meaningfully defined.

Conservation of Resources R_j

$$R_{j,k+1} = R_{j,k} - \alpha_{ij}X_{ik} - \beta_{ij}Y_{ik} - \gamma_{ij}Z_{ik} \quad (1)$$

$$R_{j,k} \geq 0 \text{ for all }_{j,k} \quad (2)$$

R_{jk}—inventory of resource j at start of year k
K_{ik}—capital inventory for industry i for year k
$g(i)$—capital gestation time for industry i
$d(i)$—depreciation for capital in industry i
$P_i(K, L)$—production capacity of industry i as a function of capital and labor
X_{ik}—level of ith production activity in year k
α_{ij}—consumption of resource j per unit activity X_{ik} (negative value implies production)
L_{ik}—labor allocated to industry i for year k
Y_{ik}—capital invested in industry i in year k
β_{ij}—consumption of resource j per unit investment activity i
Z_{ik}—level of ith consumption activity in year k
γ_{ij}—consumption of resource j per unit consumption activity i
l_k—leisure time in year k
ρ_k—population in year k
ϕ_k—total pool of labor plus leisure time for year k
λ_{jk}—marginal value (shadow price) of resource j in year k
μ_{ik}—rental value (shadow price) per unit of capital K_{ik} in year k
λ_{lk}—marginal value (shadow price) of labor in year k
Λ_{ik}—marginal value of producing capital resource K_{ik} in year k
ρ—discount rate for future utilities
$\tilde{\rho}_i$—difference between apparent discount rate for industry i and nomimal global discount rate
$U_i(Y_{ik})$—economic utility for a specified per capita level of the ith consumption activity

FIGURE 39. Definitions

Conservation† of Production Capital K

$$K_{i,k} = [1 - d(i)]\, K_{i,k-1} + Y_{i,k-g(i)} \tag{3}$$

where $g(i)$ = capital gestation time

$$K_{i,k} \geq 0 \text{ for all } _{i,k} \tag{4}$$

Production Constraints (Translog Production Function—less optimistic about the substitutability of labor for capital than the frequently used Cobb-Douglas function, $X = a(K^{\alpha}L^{1-\alpha})$

$$X_{ik} \leq P_j(K_{ik}, L_{ik}), \tag{5}$$

where

$$P_i(K_{ik}, L_{ik}) = e_{\alpha i}[\gamma_K K_{ik}^{-B_i} + \gamma_L L_{ik}^{-B_i}]^{-1/B_i} \tag{6}$$

Labor Limit Constraints

$$\sum_i L_{ik} + 1_i \leq \phi_k \tag{7}$$

Definition of Economic Objectives

The model can be viewed in two mathematically equivalent ways, either as an optimizing model which maximizes a time-integrated economic utility function, or as an economic equilibrium model which brings supply and demand into balance over the total trajectory of economic development. Viewed as an optimizing model, the objective function U is given as a function of the consumption activities Z_{ik} and in year k

$$U = \sum_k \sum_i \rho^{-pk} U_i(Z_{ik}), \tag{8}$$

where, in order to provide for higher utilities for scarce goods and services, the components of U_i of the utility function can be defined either as a simple logarithmic utility function

$$U_i(Z_{ik}) = a_i \ln (Z_{ik} - Z_i^0), \tag{9}$$

†The equation used in this model is somewhat more complicated to allow for noninteger gestation times and to reflect the integrated value over the period of time when new capital actually becomes operational.

or in the more general form of a constant-elasticity utility function.†

$$U_i(Z_{ik}) = a_i \frac{[Z_{ik} - Z_i^0]^{1-b_i-1}}{1 - b_i} \tag{10}$$

When viewed as an optimizing program, the model simply maximizes the total utility function subject to all of the previously stated constraints.

If the program is interpreted as a market equilibrium model, then the utility function $U_i(Z_{ik})$ are converted into nonlinear demand functions $D_i(Z_{ik})$ where

$$D_i(Z_{ik}) = \frac{\partial U_i(Z_{ik})}{\partial Z_{ik}} \tag{11}$$

Characteristics of the Solution

Because the two interpretations are mathematically equivalent, the same solution methodology is applicable in either case. The optimum solution involves finding sets of values X_{ik}, Y_{ik}, R_{jk}, λ_{jk}, K_{ik}, Λ_{ij}, L_{ik}, Z_{ik}, and 1_k such that the following equilibrium conditions apply.

Static Equilibrium Conditions

Demand Equilibrium

$$D_i(Z_{ik}) = \sum_j \gamma_{ij}\lambda_{jk} \tag{12}$$

Production Equilibrium

$$\lambda_{ik} = \left(\sum_j \alpha_{ij}\lambda_{jk} \right) \frac{\partial p_{ik}}{\partial L_{ik}} \tag{13}$$

Investment Equilibrium

$$\Lambda_{ik} = \sum_j \beta_{ij}\lambda_{jk} \tag{14}$$

†The equations in this section are slightly more complex in the model because utility is defined on a per capita basis to allow for population growth.

Dynamic Equilibrium Condition†

$$\Lambda_{ik'} = \sum_{k' + g(i)}^{k=\infty} e^{-\bar{\rho}_i(k-k')} \mu'_{ik} \quad (15)$$

$$\mu_{ik} = \left\{ \sum_j \alpha_{ij} \lambda_{jk} \right\} \frac{\partial P_i(K_{ik}, L_{ik})}{\partial K_{ik}} \quad (16)$$

$$\bar{\rho}_i = \rho + \tilde{\rho} + p + d_i \quad (17)$$

where p is population growth rate, d_i is the depreciation rate, and $\tilde{\rho}_i$ is the difference between effective discount rate in sector i and the general discount rate. (Note: $\tilde{\rho}_i \neq 0$ is useful in matching existing investment strategy in a nonoptimum economy.)

Solution for Static Equilibrium

The model has a capability to solve for a static equilibrium in which the values of all variables are constant over time. This capability provides a useful quick analysis approach for many problems. It is also used to provide horizon values for the dynamic optimization.

User Estimation of Structural Parameters

Most of the structural parameters for the model are determined by the basic input/output data that defines the rate of consumption and production of resources α_{ij}, β_{ij}, and γ_{ij} for each activity. The user, however, is required to estimate a few additional parameters.

For the ***production activities*** the user must provide estimates of the following parameters:

g_i the gestation time for new capital
d_i the physical depreciation rate for old capital

In addition, the user must provide an estimate of the labor vs. capital elasticity in the production function for the industry. To do this the user estimates a maximum factor by which the productivity of capital can be increased in each industry by increasing the use of labor. The program uses this information to estimate the factor B_i which controls the capital elasticity for the industry.

---

†The equation used in the model is somewhat more complicated to allow for non-integer gestation times and to reflect the integrated value over the period of time when new capital actually becomes operational.

The *population assumptions* must also be specified. Specifically, the user specifies a population growth rate ρ_i and a ratio of the available l_k to the total labor

$$\sum L_{ik}$$

in the base year for which the data are provided.

The U.S. input/output data do not specify the accumulated inventory of *production capital* in each industry. The model, however, provides an automatic estimate of this inventory which takes into account the annual capital investment and the depreciation rate. To make the estimate the user can interpret the base year either as an equilibrium state of the economy relative to population growth, or he can treat it as a non-equilibrium situation. In general, the following factors are taken into account in the estimate of the total inventory of production capital K_i in each industry:

Y_{ik}—the annual capital investment in year k for industry i

d_i—the user's estimate of the physical depreciation rate for capital in the industry

p_i—the population growth rate

g_i—the user's estimate of the rate of growth of industry i in excess of the population growth

The specific *production function* $P_i\,(K,L)$ for industry i is calculated using the investment capital K_{ik}, the labor level L_{ik}, and the production level X_{ik} for the industry in the base year together with the user's estimate of the labor capital trade-off. The resulting productivity of capital as a function of labor applied is then calculated and printed out so that adjustments in the assumptions can be made if desired.

The basic *demand coefficients* a_i for each consumption activity (see equation 10) are automatically calibrated by the program, and this process is discussed below. However, the user does have flexibility to adjust the shape of the demand curve if he wishes. Normally, as a default option, the demand elasticity is set to 1.0 and the minimum feasible consumption is set to zero. If the user wishes to experiment with a different demand elasticity b_i or a non-zero minimum of consumption Z^0 he can do so by specifying his preferred values for any of the consumption activities.

Before the model can be used to project economic behavior, the *utility function* (or economic demand function) must be calibrated so that it is compatible with the observed behavior of the economy. This calibration of the economic demand function plays a role in the DSA model that is analogous to the determination of historical statistical correlations that is used to calibrate the statistical coefficients in conventional econometric models.

Both types of models use information about past behavior to project future behavior. The traditional econometric models assume that statistical correlations in the future economy will be the same as observed in the past. The DSA economic model uses the underlying economic objectives (or economic demand) observed in past behavior as a foundation for predicting future behavior. This does not mean that the model must necessarily assume constant economic objectives. Indeed, one of the most important applications of the model is to project the response of the economy to changes in economic objectives—as for example in a military mobilization study. However, even for such studies it is important to calibrate the economic demand in the model to historical data to provide a baseline for the analysis of changes in economic demand.

The basic input/output data for the economy specify not only the coefficients α_{ij}, β_{ij}, and γ_{ij} which determine the resources produced or consumed by each activity, but also the level of operation X_i, Y_i, and Z_i for each activity. If these levels are interpreted as equilibrium levels, then the demand function can be easily calibrated. Since the resources consumed and produced in the base year are measured in terms of dollar value in that year, the shadow value λ_{jk} for each resource in the base year is simply 1.0, by definition. Using equation 12 in which both λ_{jk} and λ_{ij} are known, it is therefore possible to calculate the value of the demand functions $D(Z_{ik})$ for the per capita consumption level in the base year k. For any assumed or specified value of demand elasticity b_i and the minimum required activity Z^0_i, the demand coefficient a_i is calculated automatically for each consumption activity.

This calibration of the single-year consumption demand, however, is not sufficient to determine model behavior completely. In particular, the rate of capital investment in the model will depend on the discount rate that is applied in evaluating future values. Thus, the discount rate must also be evaluated.

Assuming that the rate of capital investment Y_{ik} observed in the base year k is close to optimum, then the marginal value for new capital Λ_{ik} for industry i in the base year k can be calculated directly, using equation 14. Moreover, the rental value of the same capital in the base year is given in equation 16. If the base year is interpreted as an economic equilibrium, then one can make use of equation 15 to calculate an effective discount rate $\bar{\rho}_i$ for the industry. Using equation 17, this effective discount rate can be decomposed into the depreciation rate, an average discount rate ρ for the economy as a whole, and an effective discount error term $\tilde{\rho}_i$ for industry i.

If the model is run using these values of ρ and $\tilde{\rho}_i$ it will reproduce the base case economy as a per capita equilibrium state.

The calibration procedures for the model, however, also permit the user to calibrate the demand function under the assumption that the base year is not really an equilibrium economy. To do this, two other parameters are provided which can be adjusted to reflect the user's interpretation of the baseline economy. Specifically, the user can specify for each production activity a physical growth rate g_i for the activity relative to the population growth rate, and a rate of change v_i for the rental value of capital equipment in the industry relative to other prices in the economy. With these assumptions the model will provide a new calibration of the economic demand. If the model is run with such a calibration, which assumes a growth in the economy in excess of the population growth rate, the model will generate a higher equilibrium and will project an economic growth toward that equilibrium.

In its present form the DSA economic model does not include the effects of technological innovation which increase productivity and produce long-term growth. Consequently, such long-term economic growth does not appear in the model calculations.

Complexity of the Model

The DSA Economic Model is currently run on the modest-sized MODCOMP Classic (Modular Computer Corporation), which has about two-thirds the CPU computational speed of a CDC 6600 and a memory of one-half million bytes. The number of equations is, patently, a function of the number of sectors, or level of disaggregation, the number of resources accounted for, the number of time periods, and the number of iterations required to achieve stable results. Currently, the model is being run with 25 sectors, 25 resources, and 40 time periods. The result is a requirement for about 8,000 equations and 55,000 16-byte words. Running time is typically about two hours. The Lagrange Multiplier Optimization Method scales approximately as the square of the problem size (compared with the cube in the case of input-output, linear programming, and related algorithms).

SYSTEM DYNAMICS

The recent review of the state-of-the-art of economic recovery modeling by Analytic Assessments Corporation, cited earlier,[2] concluded that the most hopeful technique for overcoming the limitations of past and current models is the system dynamics approach originally developed by Jay Forrester.[13] The approach was recommended because it is believed to have great flexibility and particularly because it is thought possible to include analysis of the survival and reconstitution phases, and perhaps even of the preparedness phase before a nuclear exchange. Before coming to the Validation section, therefore, we will review a system dynamics model of viability that represents the first attempt by Analytic Assessments Corporation to pursue this route.[14] This is essentially still a model of the recovery process only. Economic viability refers here to what Sidney Winter, in a pioneering study in 1963,[15] called an "inventory race" between the drawdown of inventory and inventory replenishment in the postattack period. In this concept, if the race is lost, the economy will collapse. If the replenishment rate can equal the drawdown rate before inventories of essential items are completely exhausted, the system is said to be viable and recovery will follow.

System dynamics is not a model per se but rather a paradigm that treats any socio-economic system as a set of interrelated feedback groups. Simulation of an application of the paradigm is based on a special programming language, Dynamo (Dynamo II having been used in the instant case). There are user's manuals for Dynamo II and Dynamo III, available from Pugh-Roberts Inc., Cambridge, Mass., but there is no full documentation available, as the program is treated as proprietary by its developers. The program language is said to be an order of magnitude more efficient than Fortran, for the given purpose. Essentially, the procedure is similar to that described for TAC WARRIOR in Chapter V. The system is described in terms of initial states of each variable and equations of change, or flow, per unit of time. These may be differential equations, as in TAC WARRIOR, or simple difference equations. The unit of time must be quite small and is determined by the shortest of the "first order" delays defined for the system (e.g., delivery time for supplies, processing time, gestation time for capital, even administrative or communication times). The formula used is $\Delta t = 1/2^x D$ (where D is the shortest delay time—say, one week), and sensitivity tests have shown that $x = 3(\Delta t = D/8)$ is often optimal. The program then computes, deterministically, outcomes over a specific period (or traces a path over any number of periods) by Euler's method, as described in Chapter V.

PAM4 (*Post-attack Model No. 4*)

A brief description of the Analytical Assessments Corporation model, PAM4, will provide a useful application of system dynamics to the problem under discussion. PAM4 used Winter's viability model, depicted in Figure 40,[14] as a point of departure. It focuses on the ability of the nation, under varying conditions, to achieve viability in feeding the population. The current version, PAM4, operates with four basic sectors representing the structural components of the postattack economic system: Capital Plant and

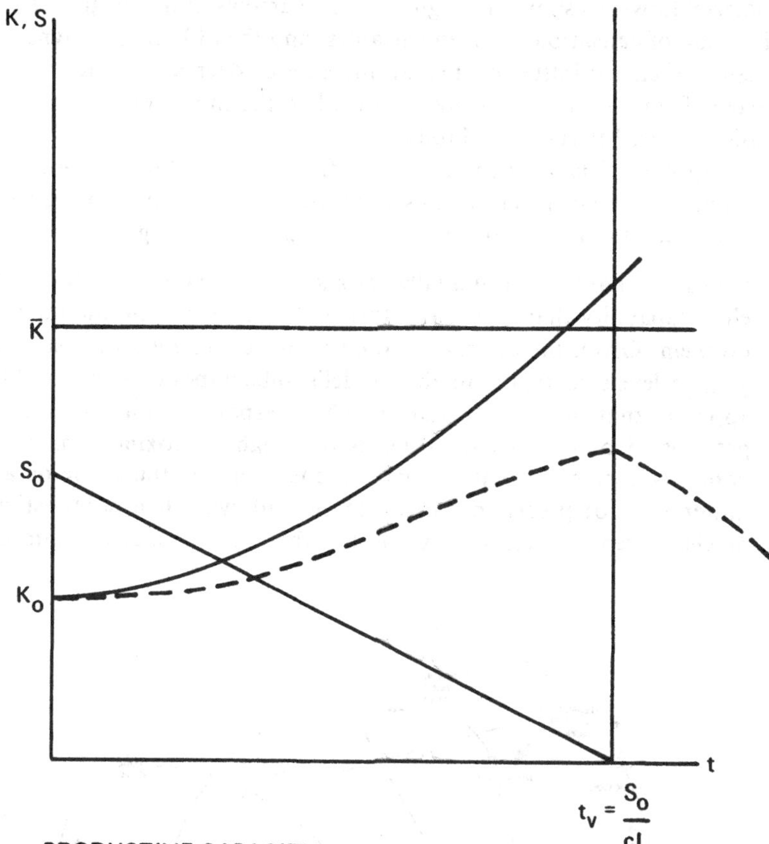

K = PRODUCTIVE CAPACITY
S = INVENTORY OF FOOD
L = LABOR
$\bar{K}$ = PRODUCTIVE CAPACITY REQUIRED FOR VIABILITY
S_o = FOOD INVENTORY AT END OF SURVIVAL PERIOD
K_o = PRODUCTIVE CAPACITY AT END OF SURVIVAL PERIOD
t_v = TIME OF DEPLETION OF FOOD INVENTORY
S_o/cL = RATIO OF FOOD STOCK TO FOOD REQUIREMENTS PER PERIOD

FIGURE 40. Success and Failure in Achieving Viability

Equipment; Intermediate Products; Labor; and Food Supplies. The Food Supplies sector is further disaggregated into Production, Transportation, and Distribution (called Matched Food Supply) subcomponents. These sectors and subsectors are interrelated through interlocking feedback groups that depict the interaction of these sectors in terms of information and material flows, as shown in Figure 41. The arrows in the diagram show the direction of causation between variables, and the (+) and (−) signs, the direction of causal influence, i.e., an increase or decrease in the variable affected. Figure 42 indicates greater detail in the most complex of the four basic sectors, Intermediate Products.

Categories of data for the states and flows are listed in an Appendix, but the sources are not given. In this connection, it is worth quoting from the chapter on "Approach" by Hill and Gardner concerning

> the use of "hard" data. Actually, data are less a requirement for a modeling capability than a general strategy for approaching the modeling problem. Given the approach noted above concerning the use of aggregate levels to represent the model's subcomponents, it should be apparent that the modeling approach we espouse is one in which the patterns of outcomes cast in terms of rough approximations are favored over precise point estimate results. For one thing, the data requirements for precision are enormous and even then point-estimate precision remains elusive except for the most immediate temporal

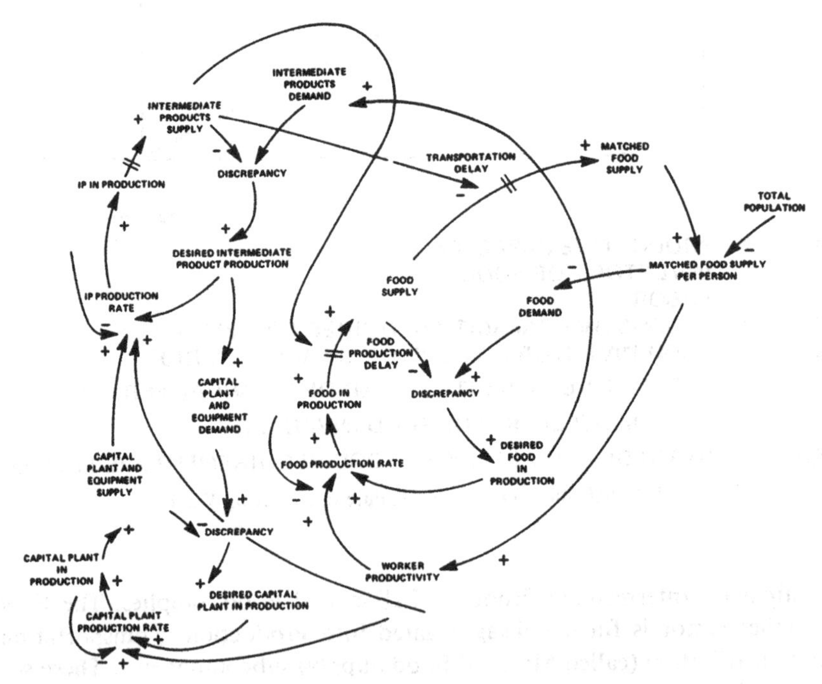

FIGURE 41. PAM4 Causal Loop Diagram: Supply Drawdown vs. Supply Replenishment

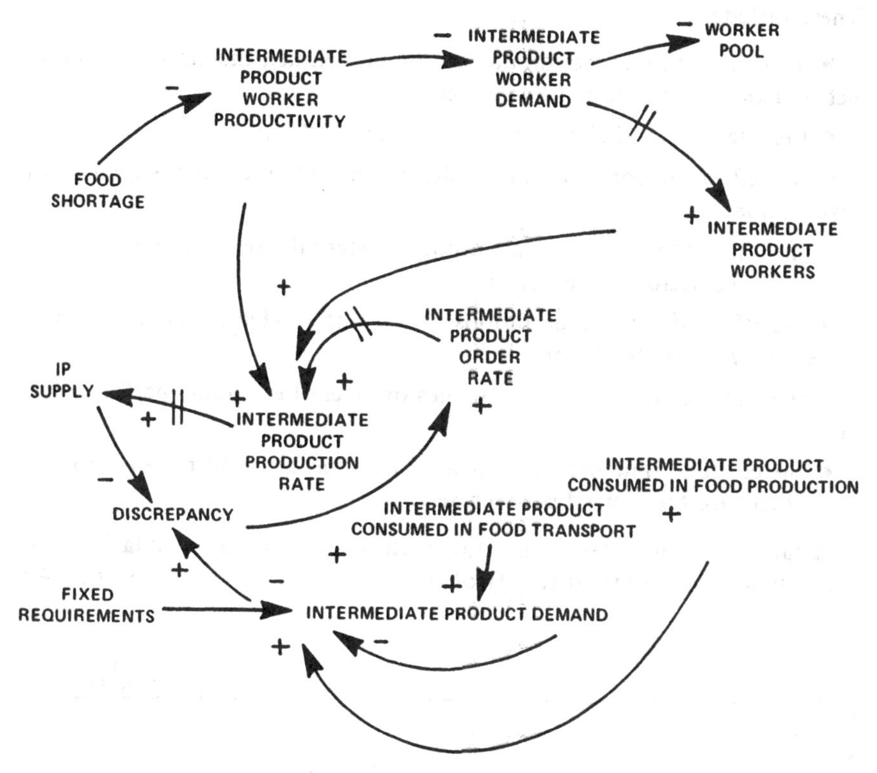

FIGURE 42. PAM4: Intermediate Products Sector

> points. Moreover, the modeler typically becomes rapidly bogged down in Herculean data gathering efforts only to discover in the long run that those data efforts were not required for each subcomponent at the level of detail needed to investigate the phenomenon of interest.[14]

The comment sounds plausible but has not been demonstrated. See [16, 17], noted below.

Model Parameters

Since the analysts were interested primarily in identifying the sources of potential instabilities affecting postattack economic viability, they set key operational parameters in PAM4 representing areas in which alternative management policies could be implemented to influence and direct the system.

These include:

- the effect of initial conditions on the model, i.e., the amount of product on hand and in the pipeline for each sector;
- the effect of food shortages on labor productivity:
- the effect of communication delays in ordering food supplies for distribution;
- the effect of various combinations of external fixed requirements that draw on each sector's inventories;
- the effect of rising expectations concerning food supplies as time progresses in the postattack period;
- the effect of labor allocation rules on overall economic performance; and
- the effect of time delays associated with locating and retraining labor as workers are transferred between sectors.

A range of values was assigned to each parameter, and simulation runs were made to test a broad variety of combinations of parameter values. An

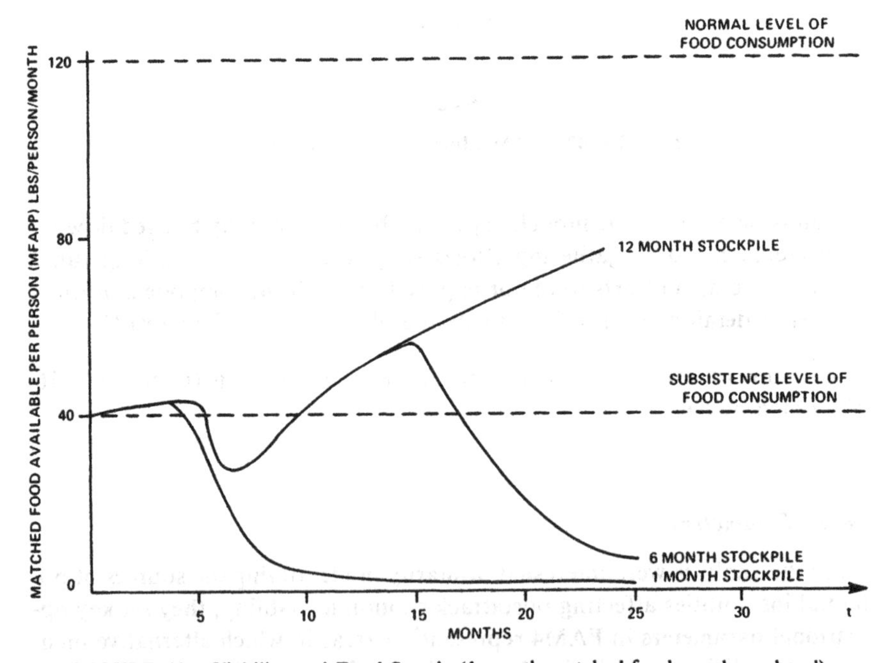

FIGURE 43. Viability and Food Supply (1 month matched food supply on hand)

example of the output from one of these runs is shown in Figure 43, where the parameter for the Matched Food Supply on hand was varied from one month's to six months' stockpile. One-third of normal (presumably prewar) food consumption is specified as subsistence level. Simulations were run for twenty-four months, "a time span that is generally regarded as the upper bound for the nation to regain viability."

The report of the study concludes that to-date "the evidence suggests that the issue of viability is greatly dependent on effective emergency preparedness policies and resource management actions. The simulation results from the PAM4 model clearly indicate that viability is not automatic even if adequate productive capacities survive; the same system can produce both viability and collapse, depending on the choice of policies and management strategies."[14] Certainly, this is an intuitively-reasonable conclusion. The model confirms the first statement but adds little to our knowledge of the validity of the second, i.e., whether viability is feasible after severe attacks.

Currently, PAM4 is run on a time-sharing basis on an IBM 370-158. Runs take a small number of minutes and repetitions for sensitivity testing are not considered a problem.

VALIDATION

This chapter has discussed the history of attempts to model recovery from nuclear war, the basic types of models that have been used, and detailed descriptions of two of the most recent attempts to improve the state-of-the-art.

As in Chapter VII on nuclear exchanges, there can be no question of testing the models against experience, despite the great contribution that the study of historical experience in recovery from wars and major disasters can make in guiding the formulation of the problem. We can, however, examine several types of validity questions to assist us in appraising the credibility of the available models.

Input Data

All of the models depend of necessity on prewar data on the economies of the United States and/or the Soviet Union, primarily in the form of interindustry, or input-output, tables. The problem is, as most analysts recognize, that in a severely—drastically—damaged postwar society, the economic relationships underlying the prewar data will change radically. In particular, input-output coefficients may have virtually no relevance. We know from wartime experience that substitution of different inputs will be widespread. Because necessity is in these circumstances truly the mother of invention, these substitutions, and the degree and rate of change in the coefficients, are unpredictable. During World War II, Germany operated virtually entirely on synthetic fuels and rubber. The United States, then the largest petroleum producer in the world, did not develop synthetic fuels, but did move rapidly into the development and use of synthetic rubber, even though as late as 1939 it was stockpiling natural rubber and had not made the basic decision to pursue the already-existing technologies for synthetic production. There were many other examples. We need only note that, of the original 1939 list of strategic materials (defined as essential and having uncertain lines of supply from overseas), most ceased to be critical sometime during the war. Cinchona bark, for example, was on the list because it was the only known source of quinine for the curing of malaria—an inevitable problem if we had to fight the Japanese in the tropics; by war's end, synthetic atabrine had become the drug of choice for the treatment of malaria. The cases can be multiplied. It is important to note, moreover, that substitution can be regarded as a subclass of a much larger set of solutions in adversity, namely, conservation. Fuel inputs are lowered by lowering space temperatures, driving less and driving slower; tin is conserved by making tin cans with thinner coatings of tin; materials are conserved by lightening structures (e.g., machine tool bases) and limiting the numbers of models; scrap metal use can be vastly increased (and scrap metal is in abundant supply in a bombed-out economy!); etc.; etc. Adjustments can be made for such potential conservation measures, but they are necessarily subjective and there is no real way to establish the validity of the adjustment factors. The analyst can never fully anticipate all of the conservation measures that would take place in a postattack economy (which resembles a wartime economy in many respects).

All postattack modeling is also dependent on damage functions calculated in nuclear war exchange models (unless they are simply postulated), and we have already seen in the last chapter the serious deficiencies in these calculations.

Postattack Policy Changes

Every analyst realizes that economic policies of individuals, firms, and governments will be different in a postattack environment. As in the above case of conservation policies, the broader set of questions about demand

and priorities is again a matter of highly-subjective prediction. The DSA Economic Model was presented because it takes a highly innovative approach to this problem, attempting to construct objective predictions of demand changes by the use of shadow prices that reflect changed relative availabilities, or supply patterns. The line of inquiry appears promising, but examination of the model did reveal the necessity for subjective judgements at many points in the analysis, and the production functions of that model, as of the others, are inexorably tied to prewar data and subjective assessments.

The system dynamics approach used in PAM4 did address directly the problem of measuring the effects of many kinds of policy changes in preparation for and during the management of recovery processes. The validity of the approach for this purpose cannot be said to have been established, however, for four basic reasons. First, the parametric handling of the policy alternatives remains both abstract and subjective. Second, the impact of these changes remains tied to production and other economic relationships based on prewar data and subjective adjustments thereof. Third, it is not clear, as in so many other cases, that the interaction of many simultaneous policy changes can be adequately understood. And, finally, the system dynamics methodology remains highly controversial, since it has been seriously challenged on the basis of its validity in earlier applications, notably those of the Urban Dynamics Model and the Club of Rome World System Dynamics Model, where predictive validity could be checked against subsequent actual trends.[16,17] See also the partial rebuttal of criticisms in [18].

Survival and Reconstitution

As noted at the outset, all models of recovery to date have assumed away the critical unknowns of whether society as well as individuals can survive a large nuclear attack and whether the governmental, societal and economic structure can be reconstituted so as to retain its legitimacy, the values for which the war was risked and fought in the first place, and the degree of control and orderliness requisite for the management of recovery. The system dynamics approach has asserted but not yet demonstrated that it can plug in "soft variables" that will measure performance in the survival and reconstitution phases, but it is not clear what could really be used for input data.

Several observations need to be made here. First, no one disputes that prewar planning can mitigate potential problems of survival and reconstitution by many means. These include direct defenses to reduce the actual damage; stockpiling supplies (quite possibly including a much broader range of goods than simply raw materials); training personnel for essential survival functions e.g., for the measurement and prediction of fallout, and providing them with widely-distributed prestocked equipment, in this case Geiger counters and dosimeters; greater protection for key personnel, from the President on down, and for the survivability of communications, data bases, and other facilities for their use; and so on. Thus, it is clear that models, such as PAM4, that parameterize such preparatory measures will show that they improve the prospect for survival and recovery. It is not clear, however, that they can provide sufficiently accurate or realistic assessments to aid the decisionmaker in making choices as to how much it is worth investing in such preparatory measures.

There is a further great danger that the modeling attempts, focusing as they do on very large attacks, i.e., very-high-damage situations, may, to the extent that they may appear adequately to measure the value of preparatory steps, price such steps out of the market when less perfect and extensive preparations might still have great value in less stringent scenarios. On the one hand, it is true that only very large attacks seriously call in question the ability of the nation to recover. On the other hand, nobody can truly weigh the "probability" of such attacks as compared to lesser contingencies. Such contingencies might include limited nuclear wars in which termination of at least the nuclear exchanges might be negotiated after, say, purely counterforce attacks on ICBMs, or on ICBMs, bomber bases and SLBM ports. Civil defense, stockpiling, etc., might be of great value in such cases, including the possibility of extended conventional conflict before and/or after the limited nuclear exchange. Moreover, and potentially most important, is the role such measures might play in the deterrence of nuclear war, which might be called the zero stage in nuclear escalation.[19]

Face Validity

Finally, we come to the ultimate test of validity, or credibility, of modeling: its apparent logic and reasonableness and its acceptance by users. The importance with which the problem is viewed is attested by the long and ongoing support of study and model development in this area, principally by the Department of Defense and the Federal Emergency Management Agency (FEMA) and its predecessor civil defense and Federal preparedness agencies. But no model, to date, appears to have been accepted by either of these principal clients as providing a definitive or even very useful policy making tool.

CONCLUSION

The relative outcomes determined by recovery models, and in particular at this stage by the DSA Economic Model, may be useful for some planning purposes, where the outcomes differ by large magnitudes, say a factor of two or more. We have in mind particularly decisions about changing targeting policy. The models should provide some guidances as to the effectiveness of particular choices of economic targets or target sets, if the criterion of the targeteer is that suggested by Secretary of Defense Rumsfeld in the *Defense Department Annual Report for Fiscal Year 1979* that "an important objective of the assured retaliation mission should be to retard significantly the ability of the USSR to recover from a nuclear exchange and regain the status of a 20th-century military and industrial power more rapidly than the United States."[20]

There may also be some limited utility of the models for the selection of alternative preparedness policies.

Nevertheless, we are forced to conclude that at the present stage recovery modeling is not based on an adequate statement of the problem. It has not addressed the analysis of the preconditions of recovery, namely, survival and reconstitution. And it has not adequately addressed the question of the national objectives, the achievement of which it is designed to measure. It has not yet found ways to solve the problem of analyzing a postattack world on the basis of preattack economic data. And finally, it does appear to have adequately addressed the problem of the interrelationship between different kinds of targets and different kinds of war scenarios. For example, while the impact of the targeting of leadership and command control facilities may be subsumed under the above-mentioned problem of the reconstitution phase, the targeting of military assets (in particular, nonnuclear forces) does not appear to have been considered. If sufficient economic assets survived for what appear to be good prospects for recovery, but the survival of military assets on the two sides were such that the postwar balance of power and of alliances were greatly altered, power might be used by one side to deny recovery possibilities to the other. For example, the Soviet Union might be able to co-opt West European resources for its recovery while interdicting U.S. imports and perhaps even extorting reparations exports.

Questions

1. What is a suitable FOM for economic recovery? Is GNP relevant? Total or per capita? Is standard of living (a per capita measure) better? How about ability to support postwar power position and political aims?

2. What is the relationship between survival, reconstitution and recovery? Recovery is presumeably not possible without survival; is survival worthwhile without recovery? Without reconstitution? Discuss.
3. Is recovery analysis on the basis of subjective estimates of postwar economic relationships better than no analysis?
4. What steps might be taken in the future to mitigate dependence on subjective estimates?
5. Can recovery analysis have any meaning if nuclear exchange models are not satisfactory?

References

1. Leontief, Wassily, ***Structure of the American Economy, 1919–1939,*** Harvard University Press, Cambridge, Massachusetts, 1941.
2. Feinberg, Abraham, "Civil Preparedness and Post-Attack U.S. Economic Recovery: A State-of-the-Art Assessment and Selected Annotated Bibliography," Marina Del Ray, California: Analytical Assessments Corporation, October 1979, Report AAC-TR-9204/79 for the Federal Emergency Management Agency.
3. Mill, John Stuart, ***Principles of Political Economy,*** 1848, cited by Feinberg [2] p. 1-1, from the 5th London edition of ***Principles of Political Economy,*** (New York: D. Appleton and Company, 1920, Volume I), pp. 108–109.
4. "Input-Output Structure of the U.S. Economy, 1972," ***Survey of Current Business,*** U.S. Department of Commerce/Bureau of Economic Analysis, April 1979, Volume 59, Number 2.
5. "Dollar-Value Tables for the 1972 Input-Output Study," ***Survey of Current Business,*** Volume 59, Number 4.
6. M. K. Wood, "PARM—An Economic Programming Model," ***Management Science,*** Volume II, Number 7, May 1965, pp. 619–680, and ***PARM System Manual,*** Volume I-A,B, "PARM: An Introduction to the Postattack Recovery Planning System," by John Dewitt Norton, National Planning Association, NCR Technical Report Number 2, August 1964.
7. Dolins, L. P., ***The IDA Civil Defense Economic Model: An Interim Summary,*** Institute for Defense Analysis (HQ 70-11362), Arlington, Virginia, March 1970.
8. Wetzler, Elliot, ***The Structure of the IDA Civil Defense Model,*** Institute for Defense Analysis (HQ 70-11741), Arlington, Virginia, August 1970 (AD 717 098).
9. Green, D. W., ***et al.***, "The SRI-WEFA Soviet Econometric Model: Phase Three Documentation," Volume I, SRI International, Strategic Studies Center, Arlington, Virginia, May 1977 (AD A043 287).
10. "The Dynamic Post-Attack Economic Model: A New Analytical Approach," Final Report, Decision-Science Applications, Inc., Arlington, Virginia, Report Number 2, September 1978.
11. Pugh, G. E., Krupp, J. C., and Galiano, R. J., "Status of the DSA Economic Recovery Model," Draft Final Technical Report, Decision-Science Applications, Inc., December 1979.
12. Pugh, G. E., "Mathematical Formulation of DSA Economic Model," DSA No. 204, Decision-Science Applications, Inc., February 1980.
13. Forrester, Jay, ***Industrial Dynamics,*** Cambridge, M.I.T. Press, 1961, ***Urban Dynamics,*** Cambridge, M.I.T. Press, 1969, and ***World Dynamics,*** Wright Allen Press, 1971.
14. Hill, Gary A. and Gardner, Peter C., ***Managing the U.S. Economy in a Post-Attack Environment: A System Dynamics Model of Viability,*** Analytical Assessments Corporation, AAC-TR-9205/79, November 1979.
15. Winter, S. G., Jr., ***Economic Viability after Thermonuclear War: The Limits of Feasible Production,*** The Rand Corporation, RM-3436-PR, September 1963.

16. Pugh, Robert, *Evaluation of Policy Simulation Models: A Conceptual Approach and Case Study*, (Washington, D.C.: Informational Resources Press, 1977).
17. Senge, Peter M., "Statistical Estimation of Feedback Models," *Simulation*, Alfred P. Sloan School of Management, Massachusetts Institute of Technology, Cambridge, Massachusetts, 1977.
18. Forrester, Jay W. and Senge, Peter M., "Tests for Building Confidence in System Dynamics Models," Systems Dynamics Group, Alfred P. Sloan School of Management, Massachusetts Institute of Technology, Cambridge, Massachusetts, 1978.
19. Hoeber, Francis P., "Civil Emergency Preparedness if Deterrence Fails," *Comparative Strategy*, Volume 1, Number 3, 1979.
20. Rumsfeld, Donald H., *Annual Defense Department Report FY 1978*, January 17, 1977.

CHAPTER IX

Epilogue

We have addressed seven examples of military modeling, in increasing order of complexity and difficulty, as noted in the Introduction. Such a small number of cases can in no sense be considered representative of the enormous and geometrically increasing number of models being developed and used across the spectrum of military planning and operations. Rather, the objective has been to treat these few cases in sufficient depth to illustrate the kinds of conceptual, analytical and data problems encountered in the practice of military modeling. The cases have perhaps been at least as illustrative of the limitations as of the accomplishments of modeling. Certainly, they have demonstrated the generalizations of the Introduction.

Chapters III to VIII, and even to some extent Chapter II, on costing, bore out the introductory remarks about the skepticism with which one should treat absolute numbers in solutions or outcomes, the much greater confidence one can usually have in relative comparisons, and the very general possibility of learning about processes and functions by the exercise of modeling and the exercising of models. The cases also substantiated the difficulties of selecting meaningful FOMs, or measures of effectiveness, and the problems of decision-making in the face of multiple relevant FOMs.

But most disturbingly of all, the cases give proof of the two greatest, and related, dilemmas of modeling. First, there is the sheer mass of data inputs and outputs, as both the scope and completeness of modeling increase, with the attendant danger for modelers and users of drowning in data before reaching the shores of increased understanding (see especially Chapter IV). Second, there is the conceptual difficulty of modeling military problems of broad scope, epitomized in Chapters III to VIII.

To let the reader in on a secret, it has been an underlying purpose of this volume to communicate a healthy skepticism about what modeling can accomplish while at the same time not discouraging the would-be practitioner but rather stimulating him to join the good fight, to seek ways to pass Scylla and Charybdis safely and reach a land of higher-level and more instructive modeling. To this end, it is worth asking what the foreseeable future holds for modeling. Will the present inadequacies be remedied? In particular, what new tools are being developed that will permit the building of not just bigger but also better models? The directions for the future efforts of modelers have already, it is hoped, been made clear, especially in the matters of modularization, hierarchical structure, and data management. We will review briefly some trends in computer hardware and software that may facilitate progress in these matters.

COMPUTER HARDWARD

Perhaps the single most significant advance in computer technology has resulted from the impact of integrated circuits on reliability, size, and cost of computer central processors. We are at the threshold of an era when MTBF will no longer impose restraints on length or running time of computer programs, and the iterative (numerical) solution of extremely difficult problems will thereby become possible. Modeling large-scale military problems where running time and, in some cases, capacity of direct access memory for the program and input data have imposed limitations even on current generation large-scale digital computers (CDC Cyber 70 series, IBM 7000 series, UNIVAC 1100 series) will be greatly facilitated.

In the future, the rapid advance in mass memory devices will cause a revolution of its own. Programmers have to date been seriously restrained by capacity of data storage, both direct and off-line. As a result, programs have become excessively convoluted, complex, and difficult to understand, primarily because of the need for conservation of limited data storage. In fact, most of the instructions in existing programs have to do with searching files for required data and restoring it to storage after use. We can expect to see startling simplification of programs to the point where they will be much more transparent to the operator, and much easier to manipulate, change, and document.

Great progress is being made in the area of computer netting, perhaps better described as interactive use of computers. This, coupled with developments in mini-computers, computer-interactive terminal equipment, and wide-band data transmission, opens new horizons for many kinds of data processing, including military modeling applications. The technical success of the ARPANET, an experimental computer network used now principally for scientific applications, is illustrative of the potential of time-shared use of large-scale machines from multiple terminals and also of the rapid transfer of data in the net. Several interesting possibilities for the user of military models are already apparent from these advances, and some of them are being utilized now by the Services. These possibilities demand structured programming, discussed below.

INNOVATIVE SOFTWARE AND ITS APPLICATION TO MODELING

One of the important and comparatively recent developments in programming, aspects of which were illustrated in Chapters III to VI, is called structured programming, or top-down structured programming. Generally, this is understood to mean that a complex problem is modeled in independent, stand-alone subroutines or modules which are driven by an executive routine that is itself comparatively simple and flexible. This modeling approach takes advantage of the developments in computer technology discussed above—specifically, the availability of inexpensive, high-capacity mass memory, with fast access to data in the memory, as well as the availability of standard software packages which improve the speed and flexibility of access to the memory and data transfer in and out of the central processor, the interactive use of computers with graphic displays, and also developments in computer netting and rapid data transfer between separate data processing facilities.

These structured programming techniques can be expected to be used increasingly in future modeling work. Some of the modeling problems discussed in other chapters of the *Casebook* may be thereby alleviated. Examples are:

—Hierarchical structures used for large-scale combat models, in order to preserve precision and realism in modeling tactical engagements while permitting aggregation of small unit engagement results to theater-level, demand structured programming. In this way, with a flexible executive routine, the model user has the option to examine a slice of the action with the resolution and sensitivity needed for his particular application.

—Combat models are, or should be, sensitive to the operational scenario and to the physical environment. This requirement can frequently be accommodated by changes in one or two subroutines or modules of a large-scale model while preserving the rest unchanged. Structured programming is ideally suited to this need for tailoring the model to specific scenarios.

—Frequently, models utilized in somewhat similar ways by a variety of users require access to a common data base which may be maintained and routinely updated at a single data-processing center. Examples are large-scale logistics and transportation models (Chapter III) which utilize the OSD/JCS Defense Force Planning Data Base, and life cycle cost models (Chapter II) which utilize cost data on particular classes of equipment maintained by Service development and procurement agencies. Structured programming and interactive time-sharing of computers should go far to alleviate the problem of physical transfer of large data files, as is now required.

—Instructional use of combat models in Service schools generally requires operation of the model in interrupted time sequence in order to permit student participation in the decision process. The result is in effect a computer-assisted war game where engagement and attrition subroutines play a portion of a battle, assess results, and present a new situation to the players. Many existing automated combat models, if they were reprogrammed in a set of structured subroutines, could be utilized in this fashion for instructional purposes. Such models would also require development of appropriate software routines to produce a suitable visual situation display.

IN CONCLUSION

The above trends are portentous technological developments. They should facilitate marked progress in military applications of modeling in the next decade. In the author's view, this progress will be significant and will lead to increasing user acceptance of analysts' contributions—but the present caveats should persist:

—Mushrooming data bases will become more manageable, but the temptation to generate more findings than we can assimilate and interpret will persist.

—The ability to manage increasingly large and realistic models by breaking them down into hierarchies of submodels will increase, but the conceptual difficulties of modeling activities of broad scope, such as theater combat, will remain with us. We will continue to be bedeviled by the mysteries of air-ground interactions, the synergism of many battalions, the two-dimensionality of battlefield movement, and so on.

—With so much running capacity at our disposal, we will be increasingly tempted to use probabilistic models without inquiring whether they reveal more about the process of interest than would a deterministic model, or whether any of the stochastic variables can, or are likely to, involve key events, the occurrence of which can significantly affect the outcome (e.g., an intelligence failure or an armor breakthrough, as distinguished from the random outcome of a one-on-one aircraft or tank engagement that will be repeated many times).

—No amount of data processing will compensate for errors in input data, or conceptual errors in the selection of FOMs or in the modeling of the process by which the FOMs are estimated.

Lest the above caveats be depressing, it is perhaps appropriate to close with the upbeat punch line of Portnoy's psychiatrist: "So [*said the doctor*]. Now vee may perhaps to begin. Yes?"[1]

References

1. Philip Roth, *Portnoy's Complaint* (New York: Random House, 1967), p. 274.

INDEX

9780367712464